Mathematical Magic Show

MARTIN GARDNER

Mathematical

THE MATHEMATICAL ASSOCIATION OF AMERICA
Washington, D.C. 1989

Magic Show

More Puzzles, Games,

Diversions, Illusions

and Other Mathematical Sleight-of-Mind

from Scientific American

with a Foreword by Persi Diaconis

and Ronald L. Graham,

Afterthoughts from the Author,

a new Postscript,

and a new Bibliography by Mr. Gardner,

Repartee from Readers,

and 139 Drawings and Diagrams.

Published in the United States of America by
The Mathematical Association of America

An MAA Spectrum book

This book was updated and revised from the 1977 edition
published by Alfred A. Knopf, Inc., New York.

Library of Congress Catalog Card Number 89-064-116
ISBN 0-88385

Manufactured in the United States of America

Cover and title page cartoons by John Johnson

For
Jim and Amy

Contents

Foreword

MANY READERS may not know the many dimensions of Martin Gardner's magic. To begin with, he is a superb sleight-of-hand artist and the inventor of hundreds of magic tricks. His earliest published pieces, written when he was in high school, were contributions to *The Sphinx*, an American magic periodical. Martin enjoys performing close-up magic for those fortunate enough to know him. He likes to bounce a dinner roll on the floor (it bounces back up like a rubber ball), to swallow a table knife, or to link a borrowed finger ring onto a rubber band. He has a special fondness for tricks that seem to violate topological laws.

A completely different kind of magic is Martin's ability to explain significant mathematical ideas to laymen, always in a way that leaves them eager for more. Unlike many other popularizers of mathematics, professionals enjoy his writing as much as amateurs. When asked how he manages this, he insists that the secret is precisely his lack of advanced knowledge. In college he took not a single mathematics course. It was not until 1989 that he finally coauthored his first formal paper that reported new discoveries.

Although self-taught in mathematics, Martin has influenced the lives of many professionals, including both of us. In one case he converted a runaway teen-age magician into a budding mathematician by publishing some of his mathematical magic ideas, and by later helping that same youth find his way into graduate work. In the other case, many fertile research problems have sprung from Martin's efforts to understand certain puzzles in order to create new ones.

Martin's success did not come easily. After graduating from the

University of Chicago in 1936 with a bachelor's degree in philosophy, he began his writing career as a newspaper reporter in Tulsa, and later as a writer in the University of Chicago's press relations office. After four years in the Navy during World War II, he began selling fiction to *Esquire*, moved to Manhattan, and became one of the editors of *Humpty Dumpty's Magazine*. After eight years of inventing bizarre activity features and writing stories and poems for five-to-eight-year old readers, he began his celebrated column in *Scientific American*. Before then, our informants tell us, he lived for years in small, dingy sleeping rooms, wearing frayed collars and pants with holes, often limiting his lunch to coffee and a piece of Danish.

Martin put a great deal of research effort into his *Scientific American* columns. He once told us that writing the column left him only a few working days a month. His main reason for retiring from the magazine was that he needed time to write books and articles about topics other than mathematics. He is now the author of more than forty volumes that range over such fields as science, philosophy, and literature as well as mathematics. His long out-of-print theological novel, *The Flight of Peter Fromm* was reissued in 1989 by Farrar, Straus, and Giroux. Many of his books are collections of literary essays and book reviews.

We visited Martin recently and were struck by the enthusiasm and boyish wonder with which he reacted to a magic sleight performed by one of us that he has not seen before—a curious method of false cutting of a deck of cards. At an age past seventy, he is as eager as he was in high school to master what magicians call a new and novel "move."

RONALD L. GRAHAM
AT&T Bell Laboratories
and Rutgers University
PERSI DIACONIS
Harvard University

Fall 1989

Introduction

THIS IS the eighth collection of my Mathematical Games columns that have been appearing monthly in *Scientific American* since December 1956. As in previous volumes, the columns have been corrected, updated, and enlarged to include bibliographies and valuable new material provided by loyal readers.

One of those readers, not mathematically inclined but who likes to read the columns nonetheless, has often asked: "Why can't you, as a favor to readers like me, give us a glossary of some of the terms you frequently use but seldom define?"

Okay, dear reader—here it is. The terms alphabetized below are so familiar to even the humblest mathematician that most readers of this book need give them no more than a passing glance. But if you are one of those adventuresome souls for whom most mathematics books are incomprehensible, but who for some strange reason have decided to look into this one, you may find it worthwhile to go over this brief, informal glossary before reading further.

Algorithm: A procedure for solving a problem, usually by repeated steps that are enormously boring unless a computer is doing them for you. You are applying algorithms when you multiply two big numbers, balance your checkbook, wash dishes, or mow the lawn.

Combination: A subset of a set, considered without regard to order. If the set is the alphabet, the subset CAT is the same three-object combination as CTA, ACT, TAC, and so on.

Combinatorial mathematics (or **combinatorics**): The study of arrangements of things. It is particularly concerned with finding out whether an arrangement that meets specified requirements is possible, and if so, how many such arrangements are possible.

Magic squares, for instance, are solutions to ancient combinatorial problems in number theory. Can the digits 1 through 9 be placed in a square array so that every row, column, and main diagonal has the same sum? Yes. How many ways can it be done? Only one if rotations and reflections are not counted as different. Can the nine digits be arranged so no two sums are alike and the sums are consecutive? No.

Composite number: An integer with two or more prime factors. Put another way, an integer other than 0, +1, or −1 that is not a prime. The first positive composites are 4, 6, 8, 9, 10. Is 1,234,567 a prime? No, it has just two prime factors and therefore is composite.

Counting numbers (or natural numbers): 1, 2, 3, 4, . . .

Digits: The ten numbers 0, 1, 2, 3, 4, 5, 6, 7, 8, 9 are the ten decimal digits. The two numbers 0, 1 are the binary digits; 0, 1, 2 are the ternary digits; and so on for the digits of higher base systems. A base-12 notation has twelve digits.

Diophantine equation: An equation in which the letters (unknown variables) stand for integers. Such equations are solved by "Diophantine analysis."

e: Next to pi, the most notorious transcendental number. It is the limit of $(1 + 1/n)^n$ as n increases without limit. In decimal notation its value is 2.718281828 . . . That crazy repetition of 1828 is sheer coincidence.

Integers: The counting numbers, their negatives, and zero.

Irrational numbers: Real numbers that are not integers. In decimal notation their fractions go on forever and have no repeating periods. Pi, e, and $\sqrt{2}$ are irrationals.

Modulo: When a number is said to equal n (modulo k) it means that when the number is divided by k it has a remainder of n. For example, $17 = 5$ (modulo 12) because 17 has a remainder of 5 when divided by 12.

N-space: A Euclidean space of n dimensions. A line is 1-space, a plane is 2-space, the world is in 3-space. A tesseract is a 4-space hypercube.

Non-negative integers: 0, 1, 2, 3, 4, 5, . . .

Order *n*: A way of classifying mathematical objects by labeling them with non-negative integers. A chessboard is a square array of order 8 if we count the number of cells on a side, order 9 if we count the lattice lines on a side instead of cells.

Permutation: An ordered subset of a set. If the set is the alphabet, CAT, CTA, ACT, and so on are different permutations of the same subset of three letters. Red, blue, white is a permutation of red, white, and blue.

Polyhedron: A solid figure bounded by polygons. A tetrahedron is a 4-sided polyhedron, a cube is a 6-sided one.

Prime: An integer, other than 0, +1, and −1, not evenly divisible by an integer except itself (plus or minus) and 1 (plus or minus). The first positive primes are 2, 3, 5, 7, 11, 13, 17, 19, ... Two interesting primes: 1,234,567,891 and 11,111,111,111,111,111,111,111. The largest known prime, discovered in 1985, is $2^{216,091} - 1$. It has 65,050 digits.

Rational numbers: Numbers that are integers, and fractions with integers above and below the line. In decimal notation an integer has either no decimal fraction, or a finite fraction, or a fraction with a repeating period.

Real numbers: The rational and irrational numbers. So called to contrast them with imaginary numbers, such as the square root of —1, even though the imaginary numbers are really just as real as the reals.

Reciprocal: A fraction turned upside down. The reciprocal of 2/3 is 3/2. The reciprocal of 3 (or 3/1) is 1/3. The reciprocal of 1 is 1.

Set: Any collection of things such as the real numbers, counting numbers, odd numbers, primes, the alphabet, the hairs on your head, the words on this page, members of Congress, and so on *ad nauseam*.

Singularity: The point at which something peculiar happens to an equation (or a physical process represented by the equation) when one or more variables have certain values. If you toss a ball in the air it reaches a singularity at the top of its path because at that precise moment its speed drops to zero. According to relativity theory, no spaceship can go faster than light because at such a speed the equations for length, time, and mass enter a singularity at which length goes to zero, time stops, and mass becomes infinite.

This introduction is about to enter the singularity at which it abruptly stops.

MARTIN GARDNER

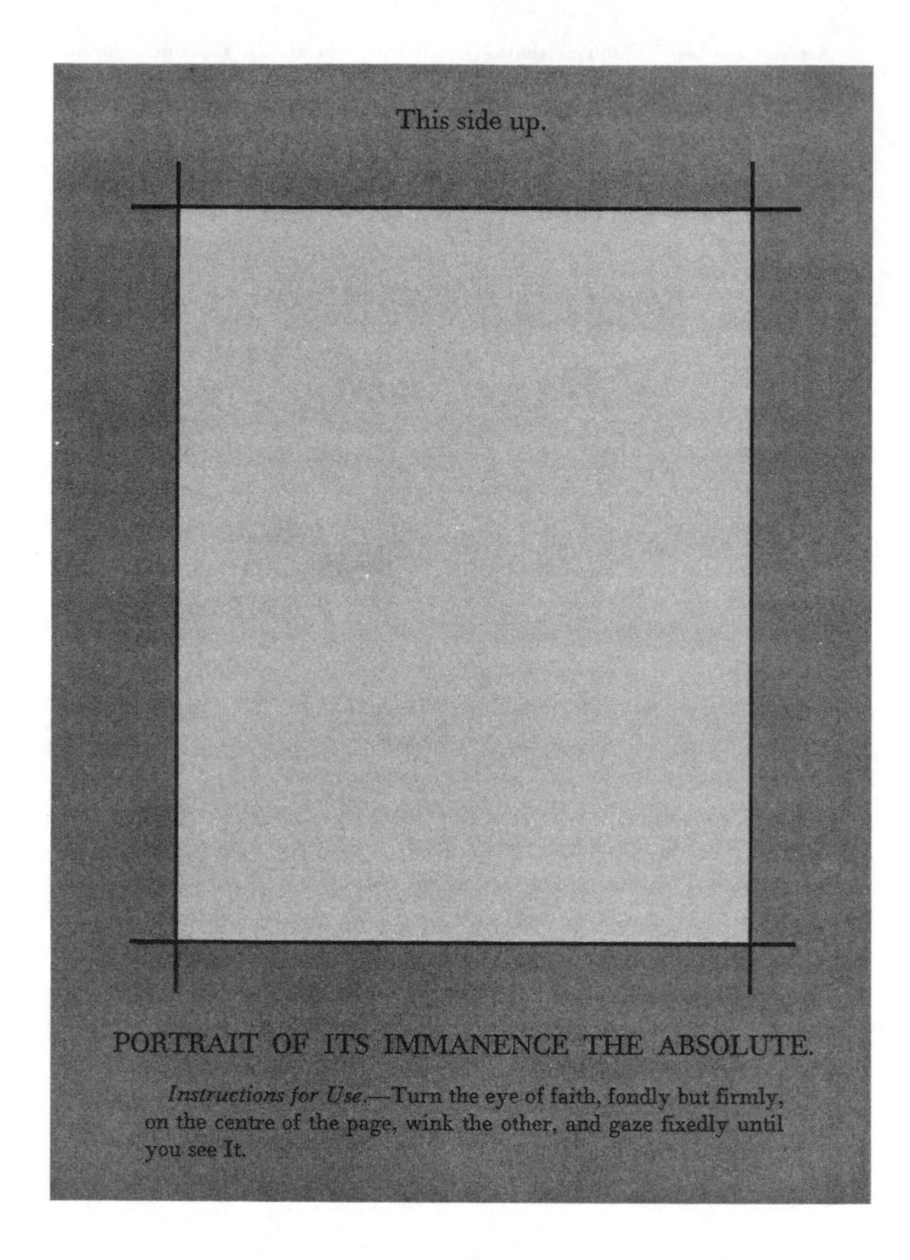

FIGURE 1

Frontispiece of Mind, *Special Christmas Number, 1901*

CHAPTER 1

Nothing

Nobody seems to know how to deal with it. (He would, of course.)

—P. L. HEATH

OUR TOPIC is nothing. By definition nothing does not exist, but the concepts we have of it certainly exist as concepts. In mathematics, science, philosophy, and everyday life it turns out to be enormously useful to have words and symbols for such concepts.

The closest a mathematician can get to nothing is by way of the null (or empty) set. It is not the same thing as nothing because it has whatever kind of existence a set has, although it is unlike all other sets. It is the only set that has no members and the only set that is a subset of every other set. From a basket of three apples you can take one apple, two apples, three apples, or no apples. To an empty basket you can, if you like, add nothing.

The null set denotes, even though it doesn't denote anything. For example, it denotes such things as the set of all square circles, the set of all even primes other than 2, and the set of all readers of this book who are chimpanzees. In general it denotes the set of all x's that satisfy any statement about x that is false for all values of x. Anything you say about a member of the null set is true, because it lacks a single member for which a statement can be false.

The null set is symbolized by $\emptyset$. It must not be confused with 0, the symbol for zero. Zero is (usually) a number that denotes the number of members of ∅. The null set denotes nothing, but 0 denotes the number of members of such sets, for example the set of apples in an empty basket. The set of these nonexisting apples is ∅, but the number of apples is 0.

A way to construct the counting numbers, discovered by the great German logician Gottlob Frege and rediscovered by Bertrand Russell, is to start with the null set and apply a few simple rules and axioms. Zero is defined as the cardinal number of elements in all sets that are equivalent to (can be put in one-to-one correspondence with) the members of the null set. After creating 0, 1 is defined as the number of members in all sets equivalent to the set whose only member is 0. Two is the number of members in all sets equivalent to the set containing 0 and 1. Three is the number of members in all sets equivalent to the set containing 0, 1, 2, and so on. In general, an integer is the number of members in all sets equivalent to the set containing all previous numbers.

There are other ways of recursively constructing numbers by beginning with nothing, each with subtle advantages and disadvantages, in large part psychological. John von Neumann, for example, shortened Frege's procedure by one step. He preferred to define 0 as the null set, 1 as the set whose sole member is the null set, 2 as the set whose members are the null set and 1, and so on.

A few years ago John Horton Conway of the University of Cambridge hit on a remarkable new way to construct numbers that also starts with the null set. He first described his technique in a photocopied typescript of thirteen pages, "All Numbers, Great and Small." It begins: "We wish to construct all numbers. Let us see how those who were good at constructing numbers have approached the problem in the past." It ends with ten open questions, of which the last is: "Is the whole structure of any use?"

Conway explained his new system to Donald E. Knuth, a computer scientist at Stanford University, when they happened to meet at lunch one day in 1972. Knuth was immediately fascinated by its possibilities and its revolutionary content. In 1973 during a week of relaxation in Oslo, Knuth wrote an introduction to Conway's method in the form of a novelette. It was issued in paperback in 1974 by Addison-Wesley, which also publishes Knuth's well-known series titled *The Art of Computer Programming*. I believe it is the only time a major mathematical discovery has been published first in a work of fiction. A later book by Conway, *On Numbers and Games*, opens with an account of his number construction, then goes on to apply the theory to the construction and analysis of two-person games. (See my *Scientific American* column, September 1976.)

Knuth's novelette, *Surreal Numbers*, is subtitled *How Two Ex-Students Turned On to Pure Mathematics and Found Total Happiness*. The book's primary aim, Knuth explains in a postscript, is not so much to teach Conway's theory as "to teach how one might go about developing such a theory." He continues: "Therefore, as the two characters in this book gradually explore and build up Conway's number system, I have recorded their false starts and frustrations as well as their good ideas. I wanted to give a reasonably faithful portrayal of the important principles, techniques, joys, passions, and philosophy of mathematics, so I wrote the story as I was actually doing the research myself."

Knuth's two ex-mathematics students, Alice and Bill (*A* and *B*), have fled from the "system" to a haven on the coast of the Indian Ocean. There they unearth a half-buried black rock carved with ancient Hebrew writing. Bill, who knows Hebrew, manages to translate the opening sentence: "In the beginning everything was void, and J. H. W. H. Conway began to create numbers." JHWH is a transliteration of how the ancient Hebrews wrote the name Jehovah. "Conway" also appears without vowels, but it was the most common English name Bill could think of that fitted the consonants.

Translation of the "Conway stone" continues: "Conway said,

'Let there be two rules which bring forth all numbers large and small. This shall be the first rule: Every number corresponds to two sets of previously created numbers, such that no member of the left set is greater than or equal to any member of the right set. And the second rule shall be this: One number is less than or equal to another number if and only if no member of the first number's left set is greater than or equal to the second number, and no member of the second number's right set is less than or equal to the first number.' And Conway examined these two rules he had made, and behold! they were very good."

The stone's text goes on to explain how on the zero day Conway created zero. He did it by placing the null set on the left and also on the right. In symbolic notation $0 = \{\emptyset \mid \emptyset\}$, where the vertical line divides the left and right sets. No member of the left $\emptyset$ is equal to or greater than a member of the right $\emptyset$ because $\emptyset$ *has* no members, so that Conway's first rule is satisfied. Applying the second rule, it is easy to show that 0 is less than or equal to 0.

On the next day, the stone reveals, Conway created the first two nonzero integers, 1 and -1. The method is simply to combine the null set with 0 in the two possible ways: $1 = \{0 \mid \emptyset\}$ and $-1 = \{\emptyset \mid 0\}$. It checks out. Minus 1 is less than but not equal to 0, and 0 is less than but not equal to 1. Now, of course, 1 and -1 and all subsequently created numbers can be plugged back into the left-right formula, and in this way all the integers are constructed. With 0 and 1 forming the left set and $\emptyset$ on the right, 2 is created. With 0, 1, and 2 on the left and $\emptyset$ on the right, 3 is created, and so on.

At this point readers might enjoy exploring a bit on their own. Jill C. Knuth's illustration for the front cover of *Surreal Numbers* shows some huge boulders shaped to symbolize $\{0 \mid 1\}$. What number does this define? And can the reader prove that $\{-1 \mid 1\} = 0$?

"Be fruitful and multiply," Conway tells the integers. By combining them, first into finite sets, then into infinite sets, the "copulation" of left-right sets continues, aided by no more than Conway's ridiculously simple rules. Out pour all the rest of the real numbers: first the integral fractions, then the irrationals. At the end of aleph-null days a big bang occurs and the universe springs into being. That, however, is not all. Taken to infinity, Conway's construction produces all of Georg Cantor's transfinite numbers, all infinitesimal numbers (they are reciprocals of infinite numbers), and infinite sets of queer new quantities such as the roots of transfinites and infinitesimals!

It is an astonishing feat of legerdemain. An empty hat rests on a table made of a few axioms of standard set theory. Conway waves two simple rules in the air, then reaches into almost nothing and pulls out an infinitely rich tapestry of numbers that form a real and closed field. Every real number is surrounded by a host of new numbers that lie closer to it than any other "real" value does. The system is truly "surreal."

"Man, that empty set sure gets around!" exclaims Bill. "I think I'll write a book called *Properties of the Empty Set.*" This notion that nothing has properties is, of course, commonplace in philosophy, science, and ordinary language. Lewis Carroll's Alice may think it nonsense when the March Hare offers her nonexistent wine, or when the White King admires her ability to see nobody on the road and wonders why nobody did not arrive ahead of the March Hare because nobody goes faster than the hare. It is easy, however, to think of instances in which nothing actually does enter human experience in a positive way.

Consider holes. An old riddle asks how much dirt is in a rectangular hole of certain dimensions. Although the hole has all the properties of a rectangular parallelepiped (corners, edges, faces with areas, volume, and so on), the answer is that there is no dirt in the hole. The various holes of our body are certainly essential to our health, sensory awareness, and pleasure. In *Dorothy and the Wizard in Oz,* the braided man, who lives on Pyramid Mountain in the earth's interior, tells Dorothy how he got there. He had been a manufacturer of holes for Swiss cheese, doughnuts, buttons, porous plasters, and other things. One day he decided to store a vast quantity of adjustable post-holes by placing them end to end in the ground, making a deep vertical shaft into which he accidentally tumbled.

The mathematical theory behind Sam Loyd's sliding-block puzzle (15 unit cubes inside a 4-by-4 box) is best explained by regarding the hole as a moving cube. It is analogous to what happens when a gold atom diffuses through lead. Bubbles of nothing in liquids, from the size of a molecule on up, can move around, rotate, collide, and rebound just like things. Negative currents are the result of free electrons jostling one another along a conductor, but holes caused by an absence of free electrons can do the same thing, producing a positive "hole current" that goes the other way.

Lao-tzu writes in Chapter 11 of *Tao Tê Ching:*

> *Thirty spokes share the wheel's hub;*
> *It is the center hole that makes it useful.*
> *Shape clay into a vessel;*
> *It is the space within that makes it useful.*
> *Cut doors and windows for a room;*
> *It is the holes which make it useful.*
> *Therefore profit comes from what is there;*
> *Usefulness from what is not there.*

Osborne Reynolds, a British engineer who died in 1912, invented an elaborate theory in which matter consists of microparticles of nothing moving through the ether the way bubbles move through liquids. His two books about the theory, *On an Inversion of Ideas as to the Structure of the Universe* and *The Sub-Mechanics of the Universe,* both published by the Cambridge University Press, were taken so seriously that W. W. Rouse Ball, writing in early editions of his *Mathematical Recreations and Essays,* called the theory "more plausible than the electron hypothesis."

Reynolds' inverted idea is less crazy than it sounds. P. A. M.

Dirac, in his famous theory that predicted the existence of antiparticles, viewed the positron (antielectron) as a hole in a continuum of negative charge. When an electron and positron collide, the electron falls into the positron hole, causing both particles to vanish.

The old concept of a "stagnant ether" has been abandoned by physicists, but in its place is not nothing. The "new ether" consists of the metric field responsible for the basic forces of nature, perhaps also for the particles. John Archibald Wheeler proposes a substratum, called superspace, of infinitely many dimensions. Occasionally a portion of it twists in such a peculiar way that it explodes, creating a universe of three spatial dimensions, changing in time, with its own set of laws and within which the field gets tied into little knots that we call "matter." On the microlevel, quantum fluctuations give space a foamlike structure in which the microholes provide space with additional properties. There is still a difference between something and nothing, but it is purely geometrical and there is nothing behind the geometry.

Empty space is like a straight line of zero curvature. Bend the line, add little bumps that ripple back and forth, and you have a universe dancing with matter and energy. Outside the utmost fringes of our expanding cosmos are (perhaps) vast regions unpenetrated by light and gravity. Beyond those regions may be other universes. Shall we say that these empty regions contain nothing, or are they still saturated with a metric of zero curvature?

Greek and medieval thinkers argued about the difference between being and nonbeing, whether there is one world or many, whether a perfect vacuum can properly be said to "exist," whether God formed the world from pure nothing or first created a substratum of matter that was what St. Augustine called *prope nihil*, or close to nothing. Exactly the same questions were and are debated by philosophers and theologians of the East. When the god or gods of an Eastern religion created the world from a great Void, did they shape nothing or something that was almost nothing? The questions may seem quaint, but change the terminology a bit and they are equivalent to present controversies.

There are endless examples from the arts—some jokes, some not—of nothing admired as something. In 1951 Ad Reinhardt, a respected American abstractionist who died in 1967, began painting all-blue and all-red canvases. A few years later he moved to the ultimate—black. His all-black five-by-five-feet pictures were exhibited in 1963 in leading galleries in New York, Paris, Los Angeles, and London. [*See Figure 2.*] Although one critic called him a charlatan (Ralph F. Colin, "Fakes and Frauds in the Art World," *Art in America*, April 1963), more eminent critics (Hilton Kramer, *The Nation*, June 22, 1963, and Harold Rosenberg, *The New Yorker*, June 15, 1963) admired his black art. An "ultimate statement of esthetic purity," was how Kramer put it (*The New York Times*, October 17, 1976) in praising an exhibit of the black paintings at the Pace Gallery.

In 1965 Reinhardt had three simultaneous shows at top Manhattan galleries: one of all-blacks, one of all-reds, one of all-blues. Prices ranged from $1,500 to $12,000. (See *Newsweek*, March 15, 1965.) For the artist's defense of his black pictures, consult *Americans, 1963*, edited by Dorothy C. Miller (The Museum of Modern Art, New York, 1963), and *Art as Art: The Selected Writings of Ad Reinhardt*, edited by Barbara Rose (Viking, 1975). (I am indebted to Thomas B. Lemann for these references.)

Since black is the absence of light, Reinhardt's black canvases come as close as possible to pictures of nothing, certainly much closer than the all-white canvases of Robert Rauschenberg and others. A *New Yorker* cartoon (September 23, 1944) by R. Taylor showed two ladies at an art exhibit, standing in front of an all-white canvas and reading from the catalogue: "During the Barcelona period he became enamored of the possibilities inherent in virgin space. With a courage born of the most profound respect for the enigma of the imponderable, he produced, at this

FIGURE 2
Ad Reinhardt: Abstract Painting, *1960–61. Oil, 60″ x 60″.*
The Museum of Modern Art

time, a series of canvases in which there exists solely an expanse of pregnant white."

I know of no piece of "minimal sculpture" that is reduced to the absolute minimum of nothing, though I expect to read any day now that a great museum has purchased such a work for many thousands of dollars. Henry Moore certainly exploited the aesthetics of holes. In 1950 Ray Bradbury received the first annual award of The Elves', Gnomes' and Little Men's Science-Fiction Chowder and Marching Society at a meeting in San Francisco. The award was an invisible little man standing on the brass plate of a polished walnut pedestal. This was not entirely nothing, says my informant, Donald Baker Moore, because there were two black shoe prints on the brass plate to indicate that the little man was actually there.

There have been many plays in which principal characters say nothing. Has anyone ever produced a play or motion picture that consists, from beginning to end, of an empty stage or screen? Some of Andy Warhol's early films come close to it, and I wouldn't be surprised to learn that the limit was actually attained by some early avant-garde playwright.

John Cage's *4′33″* is a piano composition that calls for four minutes and thirty-three seconds of total silence as the player sits frozen on the piano stool. The duration of the silence is 273 seconds. This corresponds, Cage has explained, to −273 degrees centigrade, or absolute zero, the temperature at which all molecular motion quietly stops. I have not heard *4′33″* performed, but friends who have tell me it is Cage's finest composition.

There are many outstanding instances of nothing in print: Chapters 18 and 19 of the final volume of *Tristram Shandy*, for example. Elbert Hubbard's *Essay on Silence*, containing only blank pages, was bound in brown suede and gold-stamped. I recall as a boy seeing a similar book titled *What I Know about Women*, and a Protestant fundamentalist tract called *What Must You Do to Be Lost? Poème Collectif*, by Robert Filliou, issued in Belgium in 1968, consists of sixteen blank pages.

In 1972 the Honolulu Zoo distributed a definitive monograph

called *Snakes of Hawaii: An authoritative, illustrated and complete guide to exotic species indigenous to the 50th State,* by V. Ralph Knight, Jr., B.S. A correspondent, Larry E. Morse, informs me that this entire monograph is reprinted (without credit) in *The Nothing Book.* This volume of blank pages was published in 1974, by Harmony House, in regular and deluxe editions. It sold so well that in 1975 an even more expensive (five dollars) deluxe edition was printed on fine French marble design paper and bound in leather. According to *The Village Voice* (December 30, 1974), Harmony House was threatened with legal action by a European author whose blank-paged book had been published a few years before *The Nothing Book.* He believed his copyright had been infringed, but nothing ever came of it.

Howard Lyons, a Toronto correspondent, points out that the null set has long been a favorite topic of song writers: "I ain't got nobody," "Nobody loves me," "I've got plenty of nothing," "Nobody lied when they said that I cried over you," "There ain't no sweet gal that's worth the salt of my tears," and hundreds of other lines.

Events can occur in which nothing is as startling as a thunderclap. An old joke tells of a man who slept in a lighthouse under a foghorn that boomed regularly every ten minutes. One night at 3:20 A.M., when the mechanism failed, the man leaped out of bed shouting, "What was that?" As a prank all the members of a large orchestra once stopped playing suddenly in the middle of a strident symphony, causing the conductor to fall off the podium. One afternoon in a rural section of North Dakota, where the wind blew constantly, there was a sudden cessation of wind. All the chickens fell over. A Japanese correspondent tells me that the weather bureau in Japan now issues a "no-wind warning" because an absence of wind can create damaging smog.

There are many examples that are not jokes. An absence of water can cause death. The loss of a loved one, of money, or of a reputation can push someone to suicide. The law recognizes innumerable occasions on which a failure to act is a crime. Grave consequences will follow when a man on a railroad track, in front of an approaching train and unable to decide whether to jump to the left or to the right, makes no decision. In the story "Silver Blaze," Sherlock Holmes based a famous deduction on the "curious incident" of a dog that "did nothing in the night-time."

Moments of escape from the omnipresent sound of canned music are becoming increasingly hard to obtain. Unlike cigar smoke, writes Edmund Morris in a fine essay, "Oases of Silence in a Desert of Din" (*The New York Times*, May 25, 1975), noise can't be fanned away. There is an old joke about a jukebox that offers, for a quarter, to provide three minutes of no music. Drive to the top of Pike's Peak, says Morris, "whose panorama of Colorado inspired Katharine Lee Bates to write 'America the Beautiful,' and your ears will be assailed by the twang and boom of four giant speakers—N, S, E, and W—spraying cowboy tunes into the crystal air." Even the Sistine Chapel is now wired for sound.

"At first," continues Morris, "there is something discomforting, almost frightening, about real silence. . . . You are startled by the apparent loudness of ordinary noises. . . . Gradually your ears become attuned to a delicate web of sounds, inaudible elsewhere, which George Eliot called 'that roar which lies on the other side of silence.' " Morris provides a list of a few Silent Places around the globe where one can escape not only from Muzak but from all the aural pollution that is the by-product of modern technology.

These are all examples of little pockets in which there is an absence of something. What about that monstrous dichotomy between all being—everything there is—and nothing? From the earliest times the most eminent thinkers have meditated on this ultimate split. It seems unlikely that the universe is going to vanish (although I myself once wrote a story, "Oom," about how God, weary of existing, abolished everything, including himself), but the fact that we ourselves will soon vanish is real enough. In medieval times the fear of death was mixed with a fear of eternal suffering, but since the fading of hell (albeit it is now enjoying a renaissance) this fear has been replaced by what Sören Kierkegaard called an "anguish" or "dread" over the possibility of becoming nothing.

This brings us abruptly to what Paul Edwards has called the "superultimate question." "Why," asked Leibniz, Schelling, Schopenhauer, and a hundred other philosophers, "should something exist rather than nothing?"

Obviously it is a curious question, not like any other. Large numbers of people, perhaps the majority, live out their lives without ever considering it. If someone asks them the question, they may fail to understand it and believe the questioner is crazy. Among those who understand the question, there are varied responses. Thinkers of a mystical turn of mind, the late Martin Heidegger for instance, consider it the deepest, most fundamental of all metaphysical questions, and look with contempt on all philosophers who are not equally disturbed by it. Those of a positivistic, pragmatic turn of mind consider it trivial. Since everyone agrees there is no way to answer it empirically or rationally, it is a question without cognitive content, as meaningless as asking if the number 2 is red or green. Indeed, a famous paper by Rudolf Carnap on the meaning of questions heaps scorn on a passage in which Heidegger pontificates about being and nothingness.

A third group of philosophers, including Milton K. Munitz, who wrote an entire book titled *The Mystery of Existence*, regards the question as being meaningful but insists that its significance lies solely in our inability to answer it. It may or may not have an answer, argues Munitz, but in any case the answer lies totally outside the limits of science and philosophy.

Whatever their metaphysics, those who have puzzled most over the superultimate question have left much eloquent testimony about those unexpected moments, fortunately short-lived, in which one is suddenly caught up in an overwhelming awareness of the utter mystery of why anything is. That is the terrifying emotion at the heart of Jean-Paul Sartre's great philosophical novel *Nausea.* Its red-haired protagonist, Antoine Roquentin, is haunted by the superultimate mystery. "A circle is not absurd," he reflects. "It is clearly explained by the rotation of a straight segment around one of its extremities. But neither does a circle exist." Things that do exist, such as stones and trees and himself, exist without any reason. They are just insanely *there,* bloated, obscene, gelatinous, unable not to exist. When the mood is on him, Roquentin calls it "the nausea." William James had earlier called it an "ontological wonder sickness." The monotonous days come and go, all cities look alike, nothing happens that means anything.

G. K. Chesterton is as good an example as any of the theist who, stunned by the absurdity of being, reacts in opposite fashion. Not that shifting to God the responsibility for the world's existence answers the superultimate question; far from it! One immediately wonders why God exists rather than nothing. But although none of the awe is lessened by hanging the universe on a transcendent peg, the shift can give rise to feelings of gratitude and hope that relieve the anxiety. Chesterton's existential novel *Manalive* is a splendid complement to Sartre's *Nausea.* Its protagonist, Innocent Smith, is so exhilarated by the privilege of existing that he goes about inventing whimsical ways of shocking himself into realizing that both he and the world are not nothing.

Let P. L. Heath, who had the first word in this article, also have the last. "If nothing whatsoever existed," he writes at the end of his article on nothing in *The Encyclopedia of Philosophy,* "there would be no problem and no answer, and the anxieties even of existential philosophers would be permanently laid to rest. Since they are not, there is evidently *nothing to worry about.* But that itself should be enough to keep an existentialist happy. Unless the solution be, as some have suspected, that it is not nothing that has been worrying them, but they who have been worrying it."

CHAPTER 2

More Ado About Nothing

You ain't seen nothin' yet.
—Al Jolson

The previous chapter, when it first appeared in *Scientific American* (February 1975), prompted many delightful letters on aspects of the topic I had not known about or had failed to mention. Some of this information has been worked into Chapter 1. Here is more.

Hester Elliott was the first of several readers who were reminded, by my story of the lighthouse keeper, of what some New Yorkers used to call the "Bowery El phenomenon." After the old elevated on Third Avenue was torn down, police began receiving phone calls from people who lived near the El. They were waking at regular intervals during the night, hearing strange noises, and having strong feelings of foreboding. "The schedules of the absent trains," as Ms. Elliott put it, "reappeared in the form of patterned calls on the police blotters." This is discussed, she said, by Karl Pribram in his book *Languages of the Brain* as an example of how our brain, even during sleep, keeps scanning the flow of events in the light of past expectations. It is aroused by any sharp deviation from the accustomed pattern.

Psychologist Robert B. Glassman also referred in a letter to the El example, and gave others. The human brain, he wrote, has the happy facility of forgetting, of pushing out of consciousness whatever seems irrelevant at the moment. But the irrelevant background is still perceived subliminally, and changes in this background bring it back into consciousness. Russian psychologists, he said, have found that if a human or animal listens long enough to the repeated sound of the same tone, they soon learn to ignore it. But if the same tone is then sounded in a different way, even if sounded more *softly* or more *briefly*, there is instant arousal.

Vernon Rowland, a professor of psychology at Case Western Reserve, elaborated similar points. His letter, which follows, was printed in *Scientific American*, April 1975:

SIRS:

I enjoyed Martin Gardner's essay on "nothing." John Horton Conway's rule and Gardner's analysis of "nothing" are, like all human activity, expressions of the nervous system, the study of which helps in understanding the origins and evolution of "nothing."

The brain is marvelously tuned to detect change as well as constancies in the environment. Sharp change between constancies is a perceptually or intellectually recognizable boundary. "Nothing" is "knowable" with clarity only if it is well demarcated from the "non-nothing." Even if it is vaguely bounded, nothingness cannot be treated as an absolute. This is an example of the illogicality of absolutes, because "nothing" cannot be in awareness except as it is related to (contrasted with) non-nothing.

One can observe in the brains of perceiving animals, even animals as primitive as the frog, special neurons responding specifically to spatial boundaries and to temporal boundaries. In the latter, for instance, neurons called "off" neurons, go into action when "something," say light, becomes "nothing" (darkness). "Nothing" is therefore positively signaled and is thereby endowed with existence. The late Polish neuropsychologist Jerzy Konorski pointed out the possibility that closing the eyes may activate off neurons, giving rise to "seeing" darkness and recognizing it as being different from not seeing at all.

I and others have used temporal nothingness as a food signal for cats by simply imposing 10 seconds of silence in an otherwise continuously clicking environment. Their brains show the learning of the significance of this silence in ways very similar to those for the inverse: 10 seconds of clicking presented on a continuous background of silence. "Nothing" and "something" can be treated in the same way as psychologists deal with other forms of figure-ground or stimulus-context reversal.

The nothingness of which we become aware by specific brain signals can be known only by discriminating it from other brain signals that reveal the boundaries and constancies of existing objects. This requires an act of attention. There is another form of "nothing" that is based on an attentional shift from one sense modality to another (as in the example of listening to music) or to a failure of the attentional mechanism. In certain forms of strokes the person "forgets" one part of his body and acts as if it simply does not exist, for example a man who shaves only one half of his face.

Animate systems obtain and conserve life-supporting energy by evolving mechanisms to offset or counter perturbations in their energy supply. Detecting absences ("nothings") in the energy domain had to be acquired early or survival could not have gone beyond the stage of actually living in the energy supply (protozoa in nutritious pools) rather than near it (animals that can leave the water and return).

If this pragmatic view of the biopsychological origins of "nothing" and "absence" is insufficient for trivializing the Leibnizian question ("Why should something exist rather than nothing?"), I would argue that the philosopher faces the necessity of showing that the statement "Nothing [in the absolute sense] exists" is not a self-contradiction.

The reference to my story "Oom" reminded Ms. Elliott of the following paragraph from Jorge Luis Borges' essay on John Donne's *Biathanatos* (a work which argues that Jesus committed suicide), in *Other Inquisitions, 1937–1952:*

As I reread this essay, I think of the tragic Philipp Batz, who is called Philipp Mainländer in the history of philosophy. Like me, he was an impassioned reader of Schopenhauer, under whose influence (and perhaps under the influence of the Gnostics) he imagined that we are fragments of a God who destroyed Himself at the beginning of time, because He did not wish to exist. Universal history is the obscure agony of those fragments. Mainländer was born in 1841; in 1876 he published his book Philosophy of the Redemption. *That same year he killed himself.*

Is Mainländer one of Borges' invented characters? No, he actually existed. You can read about him and his strange two-volume work in *The Encyclopedia of Philosophy*, Vol. 6, page 119.

Several readers informed me of the amusing controversy among graph theorists over whether the "null-graph" is useful. This is the graph that has no points or edges. The classic reference is a paper by Frank Harary and Ronald C. Read, "Is the Null-Graph a Pointless Concept?" (The paper was given at the Graphs and Combinatorial Conference, at George Washington University, in 1973, and appears in the conference lecture notes, published by Springer-Verlag.)

"Note that it is not a question of whether the null-graph 'exists,' " the authors write. "It is simply a question of whether there is any point in it." The authors survey the literature, give pros and cons, and finally reach no conclusion. Figure 3, reproduced from their paper, shows what the null-graph looks like.

Wesley Salmon, the philosopher of science, sent a splendid ontological argument for the existence of the null set:

I have just finished reading, with much pleasure, your column on "nothing." It reminded me of a remark made by a brilliant young philosopher at the University of Toronto, Bas van Fraassen, who, in a lecture on philosophy of mathematics, asked why there might not be a sort of ontological proof for the existence of the null set. It would begin, "By the null set we under-

FIGURE 3
The null-graph

stand that set than which none emptier can be conceived . . ." Van Fraassen is editor in chief of the Journal of Philosophical Logic. *I sent him the completion of the argument:*

"The fool hath said in his heart that there is no null set. But if that were so, then the set of all such sets would be empty, and hence, it *would be the null set. Q.E.D."*

I still do not know why he did not publish this profound result.

Frederick Mosteller, a theoretical statistician at Harvard, made the following comments on the superultimate question:

Ever since I was about fourteen years old I have been severely bothered by this question, and by and large not willing to talk to other people about it because the first few times I tried I got rather unexpected responses, mainly rather negative put-downs. It shook me up when it first occurred to me, and has bothered me again and again. I could not understand why it wasn't in the newspapers once a week. I suppose, in a sense, all references to creation are a reflection of this same issue, but it is the simplicity of the question that seems to me so scary.

When I was older I tried it once or twice on physicists and again did not get much of a response—probably talked to the wrong ones. I did mention it to John Tukey once, and he offered a rather good remark. He said something like this: contemplating the question at this time doesn't seem to be producing much information—that is, we aren't making much progress with it—and so it is hard to spend time on it. Perhaps it is not yet a profitable question.

It seems so much more reasonable to me that there should be nothing than something that I have secretly concluded for myself that quite possibly physicists will ultimately prove that, were there a system containing nothing, it would automatically create a physical universe. (Of course, I know they can't quite do this.)

CHAPTER 3

Game Theory, Guess It, Foxholes

THE THEORY

GAME THEORY, one of the most useful branches of modern mathematics, was anticipated in the early 1920's by the French mathematician Emile Borel, but it was not until 1926 that John von Neumann gave his proof of the minimax theorem, the fundamental theorem of game theory. On this cornerstone he built almost single-handedly the beautiful basic structure of game theory. His classic 1944 work, *Theory of Games and Economic Behavior*, written with the economist Oskar Morgenstern, created a tremendous stir in economic circles (see "The Theory of Games," by Oskar Morgenstern, *Scientific American*, May 1949). Since then game theory has developed into a fantastic amalgam of algebra, geometry, set theory, and topology, with applications to competitive situations in business, warfare, and politics as well as economics.

Attempts have been made to apply game theory to all kinds of other conflict situations. What is the nation's optimal strategy in the Cold War Game? Is the Golden Rule, some philosophers have asked, the best strategy for maximizing happiness payoffs in the Great Game of Life? How can a scientist best play the Induction Game against his formidable opponent Nature? Even psychiatry has not been immune. Although Eric Berne's "transactional therapy" (popularized by his best-selling *Games People Play*) makes no use of game theory mathematics, it borrows many of its terms from, and obviously has been influenced by, the game theory approach.

Most game theory work has been on what are called two-person zero-sum games. This means that the conflict is between two players (if there are more, the theory gets muddied by coalitions) and whatever one player wins the other loses. (One reason game theory is difficult to apply to international conflicts is that they are not zero-sum; a loss for the U.S.S.R. is not necessarily a gain for the United States, for example.) The main purpose of this chapter is to present an interesting two-person zero-sum card game invented by Rufus Isaacs, a game theory expert who wrote *Differential Games* (John Wiley, 1965) and is professor of applied mathematics at Johns Hopkins University. But first a quick look at some elementary game theory.

Consider this trivial game. Players A and B simultaneously extend one or two fingers, then B gives A as many dollars as there are fingers showing. The game obviously is unfair since A always wins. How, though, should A play so as to make his wins as big as possible, and how should B play so as to lose as little as possible? Most games have numerous and complicated strategies, but here each player is limited to two: he can show one finger or he can show two. The "payoff matrix" can therefore be drawn on a 2-by-2 square as shown in Figure 4, left. By convention, A's two strategies are shown on the left and B's two strategies are shown above. The cells hold the payoffs for every combination of strategies. Thus if A shows one finger and B two, the intersection cell shows a $3 payoff to A. (Payoffs are always given as payments from B to A even when the money actually goes the other way, in which case the payment to B is indicated by a minus sign.)

If A plays one finger, the least he can win is 2. If he plays two fingers, the least he can win is 3. The *largest* of these lows (the 3 at lower left) is called the maxmin (after maximum of

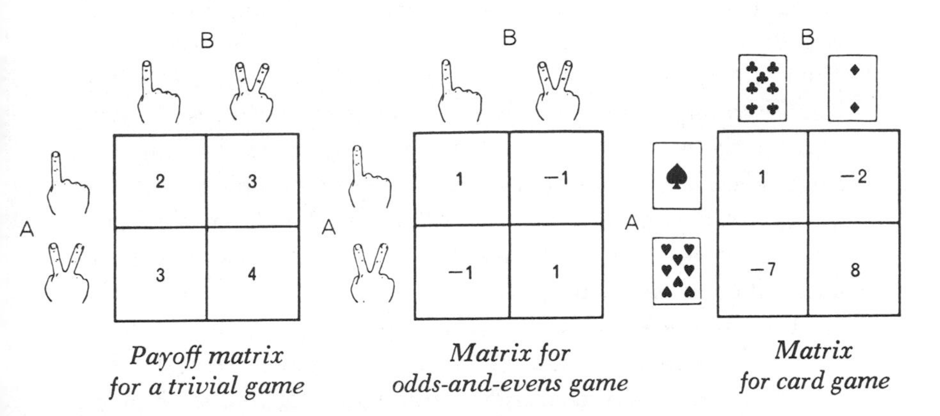

Payoff matrix for a trivial game

Matrix for odds-and-evens game

Matrix for card game

FIGURE 4

the minima). If B plays one finger, the most he can lose is 3. If he plays two fingers, the most he can lose is 4. The least of these highs (again the 3 at lower left) is called the minmax (minimum of the maxima). If the cell that holds the minmax is also the cell that holds the maxmin, as it is in this case, the cell is said to contain the game's "saddle point" and the game is "strictly determined."

Each player's best strategy is to play a strategy that includes the saddle point. A maximizes his gain by always showing two fingers; B minimizes his loss by always showing one. If both play their best, the payoff each time will be \$3 to A. This is called the "value" of the game. As long as either player uses his optimal strategy he is sure to receive a payoff equal to or better than the game's value. If he plays a nonoptimal strategy, there is always an opposing strategy that will give him a poorer payoff than the value. In this case the game is of course so trivial that both optimal strategies are intuitively obvious.

Not all games are strictly determined. If we turn the finger game into "odds and evens" (equivalent to the game of matching pennies), the payoff matrix becomes the one shown in Figure 4, middle. When fingers match, A wins \$1; when they do not match, B wins $1. Since A's maxmin is −1 and B's minmax is 1, it is clear there is no saddle point. Consequently neither player finds one strategy better than the other. It would be foolish, for example, for A to adopt the strategy of always showing two fingers because B could win every time by showing one finger. To play optimally each player must mix his two strategies in certain proportions. Ascertaining the optimal proportions can be difficult, but here the symmetry of this simple game makes it obvious that they are 1 : 1.

This introduces an all-important aspect of game theory: to be effective the mixing must be done by a randomizing device. It is easy to see why nonrandom mixing is dangerous. Suppose A mixes by alternating one and two fingers. B catches on and plays to win every time. A can adopt a subtler mixing pattern but there is always the chance that B will discover it. If he tries to randomize in his head, unconscious biases creep in. When Claude E. Shannon, the founder of information theory, was at the Bell Telephone Laboratories, he and his colleague D. W. Hagelbarger each built a penny-matching computer that consistently won against human players when they made their own choices by pressing one of two buttons. The computer analyzed its opponent's plays, detected nonrandom patterns, and played accordingly. Because the two machines used different methods of analyzing plays, they were pitted against each other "to the accompaniment," Shannon disclosed, "of small side bets and loud cheering" (see "Science and the Citizens," *Scientific American*, July 1954). The only way someone playing against such a machine can keep his average payoff down to zero is to use a randomizer—for example, flipping a penny each time to decide which button to push.

The game matrix shown in Figure 4, right, provides an amusing instance of a game with a far from obvious mixed strategy. Player A holds a double-faced playing card made by pasting a black ace back to back to a red eight. Player B has a similar double card: a red two pasted to a black seven. Each chooses a side of his card and simultaneously shows it to the other. A wins if the colors match, B if they fail to match. In every case the payoff in dollars is equal to the value of the winner's card.

The game *looks* fair (has a value of zero) because the sum of what A can win ($8 + 1 = 9$) is the same as the sum of what B can win ($2 + 7 = 9$). Actually the game is biased in favor of B, who can win an average of $1 every three games if he mixes his two strategies properly. Since 8 and 1, in one diagonal, are each larger than either of the other two payoffs, we know at once that there is no saddle point. (A 2-by-2 game has a saddle point if and only if the two numbers of either diagonal are *not* both higher than either of the other two numbers.) Each player, therefore, must mix his strategies.

Without justifying the procedure, I shall describe one way to calculate the mixture for each player. Consider A's top-row strategy. Take the second number from the first: $1 - (-2) = 3$. Do the same with the second row: $-7 - 8 = -15$. Form a fraction (ignoring any minus signs) by putting the last number above the first: 15/3, which simplifies to 5/1. A's best strategy is to mix in the proportions 5 : 1, that is, to show his ace five times for every time he shows his seven. A die provides a convenient randomizer. He can show his ace when he rolls 1, 2, 3, 4, or 5, his seven when he rolls 6. The randomizer's advice must, of course, be concealed from his opponent, who otherwise would know how to respond.

B's best strategy is similarly obtained by taking the bottom numbers from the top. The first column yields 8, the second -10. Ignoring minus signs and putting the second above the first gives 10/8, or 5/4. B's best strategy is to show his seven five times to every four times for the two. As a randomizer he can use a table of random numbers, playing the seven when the digit is 1, 2, 3, 4, or 5 and the deuce when it is 6, 7, 8, or 9.

To calculate the game's value (the average payoff to A), assume that the cells are numbered left to right, top to bottom, a, b, c, d. The value is

$$\frac{ad - bc}{a + d - b - c}.$$

The formula in this case has a value of $-1/3$. As long as A plays his best strategy, the 5 : 1 mixture, he holds his average loss per game to a third of a dollar. As long as B plays his best mixture, the 5 : 4, he ensures an average win per game of a third of a dollar. The fact that every matrix game, regardless of size or whether it has a saddle point, has a value, and that the value can be achieved by at least one optimal strategy for each player, is the famous minimax theorem first proved by von Neumann. Readers may enjoy experimenting with 2-by-2 card games of this type but using different cards, and calculating each game's value and optimal strategies.

Most two-person board games, such as chess and checkers, are played in a sequence of alternating moves that continues until either one player wins or the game is drawn. Since the number of possible sequences is vast and the number of possible strategies is astronomically vaster, the matrix is much too enormous to draw. Even as simple a game as ticktacktoe would require a matrix with tens of thousands of cells, each labeled 1, −1, or 0. If the game is finite (each player has a finite number of moves and a finite number of choices at each move) and has "perfect information" (both players know the complete state of the game at every stage before the current move), it can be proved (von Neumann was the first to do it) that the game is strictly determined. This means that there is at least one best pure strategy that always wins for the first or for the second player, or that both of the players have pure strategies that can ensure a draw.

THE GAME OF GUESS IT

ALMOST ALL card games are of the sequential-move type but with incomplete information. Indeed, the purpose of making the backs of cards identical is to conceal information. In such games the optimal strategies are mixed. This means that a player's best decision on most or all of his moves can be given only probabilistically and that the value of the game is an average of what the maximizing player will win in the long run. Poker, for instance, has a best mixed strategy, although (as in chess and checkers) it is so complicated that only simplified forms of it have been solved.

Isaacs' card game, named Guess It by his daughter Ellen, is remarkable in that it is a two-person sequential-move game of incomplete information, sufficiently complicated by bluffing to make for stimulating play, yet simple enough to allow complete analysis.

The game uses eleven playing cards with values from ace to jack, the jack counting as 11. The packet is shuffled. A card is drawn at random and placed face down in the center of the table, neither player being aware of its value. The remaining ten cards are dealt, five to each player. The object of the game is to guess the hidden card. This is done by asking questions of the form "Do you have such-and-such a card?" The other player must answer truthfully. No card may be asked about twice.

At any time, instead of an "ask" a player may end the game by a "call." This consists of naming the hidden card. The card is then turned over. If it was correctly named, the caller wins; otherwise he loses. To play well, therefore, a player must try to get as much information as he can, at the same time revealing as little as possible, until he thinks he knows enough to call. The delightful feature of the game is that each player must resort to occasional bluffing, that is, asking about a card he himself holds. If he never bluffed, then whenever he asked about a card not in his opponent's hand, the opponent would immediately know that card must be the hidden one—and would call and win. Bluffing is therefore an essential part of strategy, both for defense and for tricking the opponent into a false call.

If player A asks about a card, say the jack, and the answer is yes, both players will then know B has that card. Since it will not be asked about again, nor will it be called, the jack plays no further role in the game. B places it face up on the table.

If B does not have the jack, he answers no. This places him

in a quandary, although one that proves to be short-lived. If he thinks A is not bluffing, he calls the jack and ends the game, winning if his suspicion is correct. If he does not call it and the hidden card *is* the jack, then A (who originally asked about it) will surely call the jack on his next play, for he will know with certainty that it is the hidden card. Therefore, if A does *not* call the jack on his next play, it means he had previously bluffed and has the jack in his hand. Again, because the location of the card then becomes known to both players, it plays no further role. It is removed and placed face up on the table. In this way hands tend to grow smaller as the game progresses. After each elimination of a card the players are in effect starting a new game with fewer cards in hand.

It is impossible to give here the details of how Isaacs solved the game. The interested reader will find it explained in his article "A Card Game with Bluffing" in *The American Mathematical Monthly* (Vol. 62, February 1955, pages 99–108). I will do no more here than explain the optimal strategies and how they can be played with the aid of two spinners made with the dials shown in Figure 5. Readers are urged first to play the game without these randomizers, keeping a record of n games between players A and B. They should then play another n games with only A using the spinners, followed by a third set of n games with only B using the spinners. (If both players use randomizers, the game degenerates into a mere contest of chance.) In this way an empirical test can be made of the efficacy of the strategy.

The dials can be copied or mounted on a rectangle of stiff cardboard. Stick a pin in the center of each and over each pin put the loop end of a bobby pin. A flip of the finger sends the bobby pin spinning. The spinners must of course be kept out of your opponent's view when being used, either by turning your back when you spin them or keeping them on your lap below the edge of the table. After using them you must keep a "poker face" to avoid giving clues to what the randomizers tell you to do.

The top dial tells you when to bluff. The boldface numbers

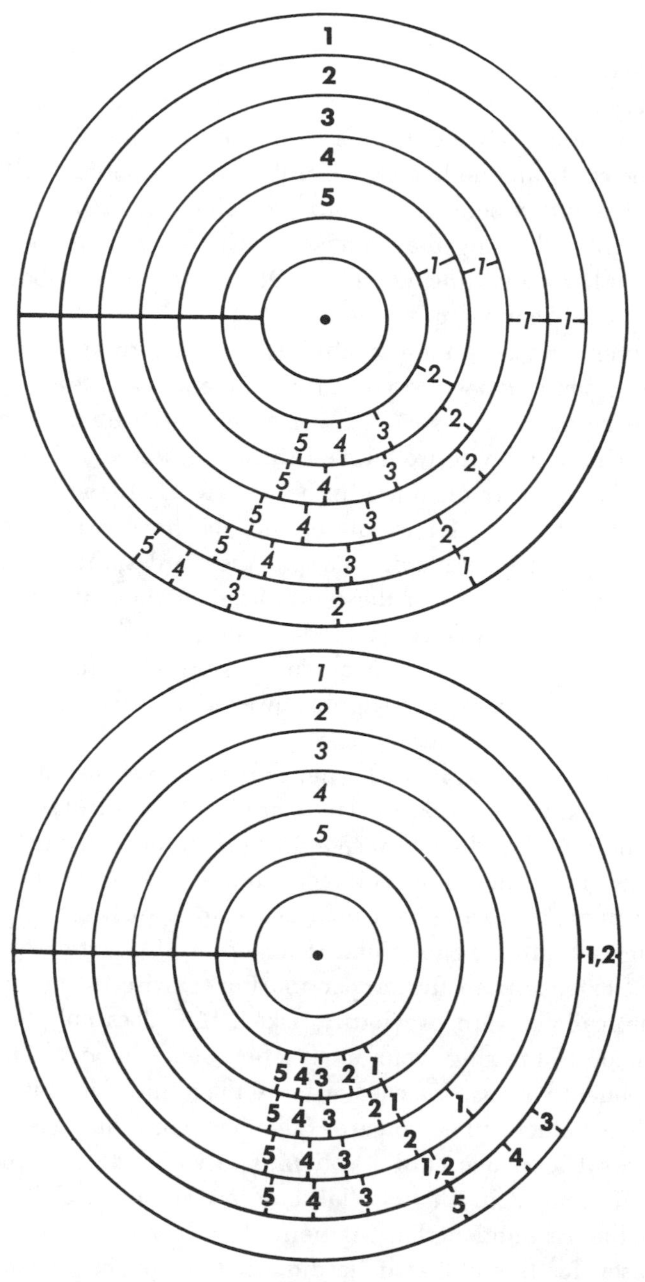

FIGURE 5

Randomizing dials for deciding when to bluff (top) and when to call (below)

give the number of cards in your hand. The other numbers scattered over the dial and attached to marks stand for the number of cards in your opponent's hand. Assume that you have three cards and he has two. Confine your attention to the ring labeled with a boldface 3. Spin the bobby pin. If it stops in the portion of the ring that extends clockwise from mark 2 to the heavy horizontal line, you bluff. Otherwise you ask about a card that could be in your opponent's hand.

In either case, asking or bluffing, pick a card at random from the possibilities open to you. If a strict empirical test of strategy is to be made, you should use a randomizer for this selection. The simplest device would be a third spinner on a circle divided into 11 equal sectors and numbered 1 to 11. If the first spinner tells you to bluff, for example, and you have two, four, seven, and eight in your hand, you spin the third spinner repeatedly until it stops on one of those numbers. Without the aid of such a spinner, simply select at random one of the four cards in your hand. The danger of your opponent's profiting from an unconscious mental bias is so slight, however, that we shall assume a third spinner is not used.

The bottom dial is used whenever you have just answered no to an ask. On this dial the rings are labeled with italic numbers to indicate that they correspond to the number of cards in your opponent's hand. The boldface numbers near the marks give the number of cards you hold. As before, pick the appropriate ring and spin the bobby pin. If it stops in the portion of the ring that extends from the proper mark clockwise to the horizontal line, call the card previously asked. If it does not stop in this portion of the ring, your next action depends on whether your opponent has just one card or more than one. If he has only one, call the other unknown card. If he has more than one (and you have at least one card), you must ask. To decide whether to bluff or not, spin the first dial, but now you must pick your ring on the assumption that his hand is reduced by one card. The reason for this is that if he did not bluff on his last ask, your "no" answer will enable him to win on his next move. You therefore play as if he were bluffing and the game were to continue, in which case the card he asked about has been taken out of the game by your "no" answer even though it is not actually placed face up on the table until after his next move.

In addition to the circumstances just explained, you call only under the following circumstances: (1) When you know the hidden card. (This occurs when you have asked without bluffing and received a "no" reply, and he has not won the game by calling on his next turn; and it occurs of course when he holds no cards.) (2) When you have no cards and he has one or more, because if you do not call, he surely will call and win on his next play. If each of you holds just one card, it is immaterial whether you call or ask; the probability of winning is 1/2 and is obtained either way. (3) When instructed to call by the second dial, as explained before.

The table in Figure 6 shows the probability of winning for the player who has the move. The number of his cards appears at the top, those of the other player on the left. At the beginning, assuming that both players use randomizers for playing their best, the first player's probability of winning is .538, or

NUMBER OF CARDS IN OPPONENT'S HAND \ NUMBER OF CARDS IN PLAYER'S HAND	1	2	3	4	5
1	.5	.667	.688	.733	.75
2	.5	.556	.625	.648	.680
3	.4	.512	.548	.597	.619
4	.375	.450	.513	.543	.581
5	.333	.423	.467	.512	.538

FIGURE 6
Chart of probabilities of winning Guess It game

slightly better than 1/2. If the payoff to the first player is $1 for each win and zero for each loss, then $.538 is the value of the game. If after each game the loser pays the winner $1, the first player will win an average of 538 games out of every 1,000. Since he receives $538 and loses $462, his profit is $76, and his average win per game is $76/1,000, or $.076. With these payoffs the game's value is a bit less than eight cents per game. If the second player does not use randomizers, the first player's chance of winning increases substantially, as should appear in an empirical test of the game.

FOXHOLES

HERE IS a simple, idealized war game that Isaacs uses to explain mixed strategies to military personnel. One player, the soldier, has a choice of hiding in any one of the five foxholes shown in Figure 7. The other player, the gunner, has a choice of firing at one of the four spots *A, B, C, D*. A shot will kill the soldier if he is in either adjacent foxhole—shot *B*, for example, is fatal if he is in foxhole 2 or 3.

"We can see the need for mixing strategies," Isaacs writes, "for the soldier might reason: 'The end holes are vulnerable to only one shot, whereas the central holes can each be hit two ways. Therefore I'll hide in one of the end holes.' Unfortunately the gunner might foresee this reasoning and fire only at *A* or *D*. If the soldier suspects that the gunner will do this, he will hide in a central hole. But now the gunner may still be one-up by guessing that the soldier will think he will think this way, therefore he aims at the center. These attempts at outthinking the opponent lead only to chaos. The only way either player can be sure of deceiving his opponent is by mixing his strategies."

Assume that the payoff is 1 if the gunner kills the soldier, 0 if he does not. The value of the game is then the same as the probability of a hit. What are the optimal strategies for each player and what is the game's value?

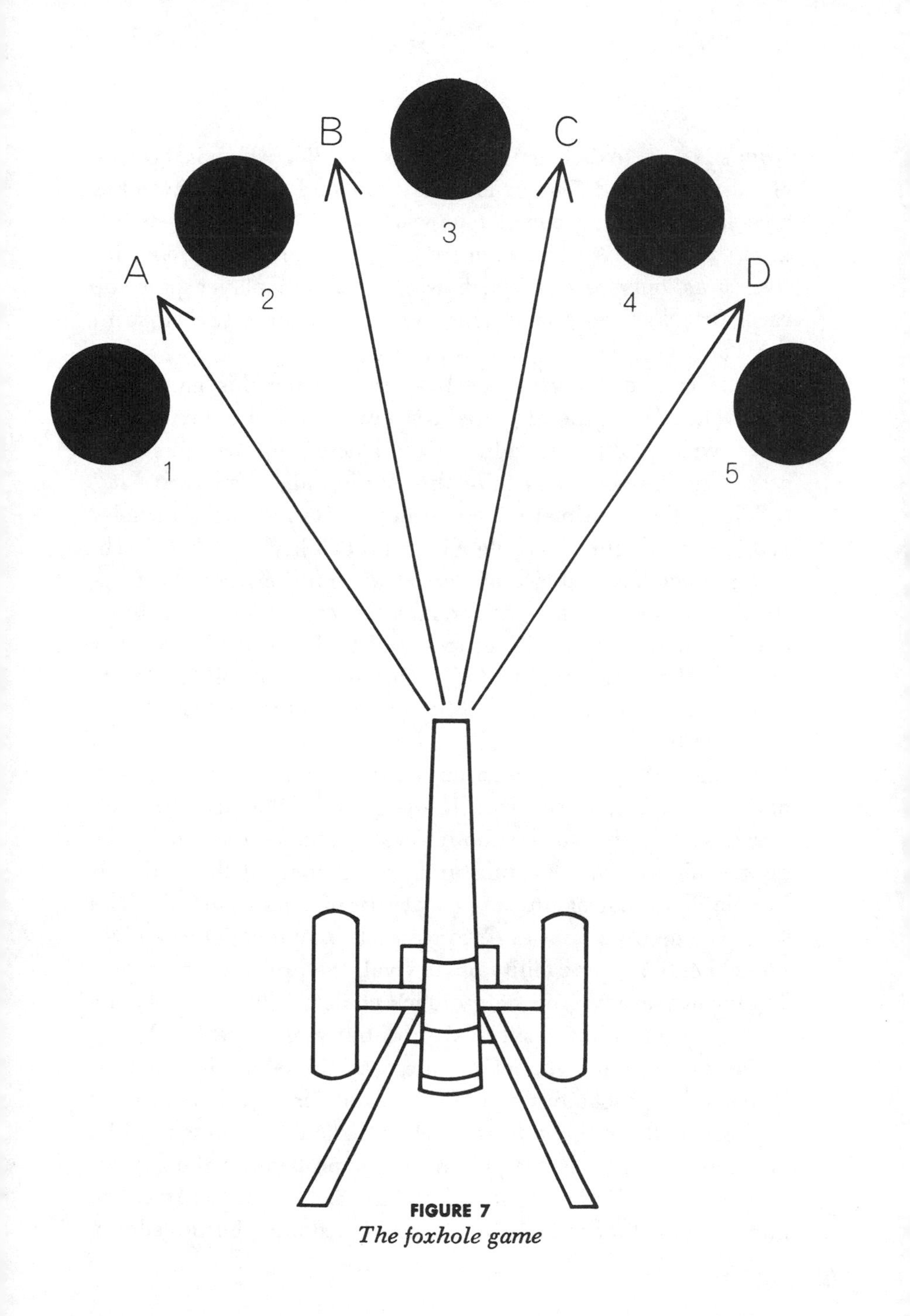

FIGURE 7
The foxhole game

ANSWERS

RUFUS ISAACS' foxhole game concerns a soldier who has a choice of hiding in one of five foxholes in a row and a gunner who has a choice of firing at one of four spots, *A*, *B*, *C*, *D*, between adjacent foxholes. An equivalent card game can be played with five cards, only one of which is an ace. One player puts the cards face down in a row. The other player picks two adjacent cards and wins if one of them is the ace.

"One can easily write a 4-by-5 matrix for this game and apply one of the general procedures described in the textbooks," Isaacs writes. "But, with a little experience, one learns in simple cases like this how to surmise the solution and then verify it."

The soldier's optimal mixed strategy is to hide only in holes 1, 3, and 5, selecting the hole with a probability of 1/3 for each. The gunner has a choice of any of an infinite number of optimal strategies. He assigns probabilities of 1/3 to *A*, 1/3 to *D*, and any pair of probabilities to *B* and *C* that add to 1/3. (For example, he could let *B* and *C* each have a probability of 1/6, or he could give one a probability of 1/3 and the other a probability of 0.)

To see that these strategies are optimal, consider first the soldier's probability of survival. If the gunner aims at *A*, the soldier has a 2/3 chance of escaping death. The same is true if the gunner aims at *D*. If he aims at *B*, he hits only if the soldier is in hole 3, so that again the probability of missing is 2/3. The same is true if he aims at *C*. Since each individual choice gives the soldier a 2/3 probability of survival, the probability remains 2/3 for any mixture of the gunner's choices. Thus the soldier's strategy ensures him a survival probability of at least 2/3.

Consider now the gunner's strategy. If the soldier is in hole 1, he has a hit probability of 1/3. If the soldier is in hole 2, he is hit only if the gunner fires at *A* or *B*, and consequently the probability of a hit is 1/3 plus whatever probability the gunner assigned to *B*. If the soldier is in hole 3, he is hit only if the gunner fires at *B* or *C*, to which are assigned probabilities adding to 1/3. Therefore the probability of a hit here is 1/3. If the soldier is in hole 4, the probability of a hit is 1/3 plus the probability assigned to *C*. If he is in hole 5, the probability is 1/3. Thus the gunner's strategy guarantees him a probability of at least 1/3.

Assuming a payoff of 1 to the gunner if he kills the soldier, 0 if he doesn't, the value of the game is 1/3. The gunner has an infinite number of strategies that guarantee him a hit probability of at least 1/3. It is possible he could do better against a stupid opponent, but against good opposition he can hope for no more because, as we have seen, the soldier has a strategy that keeps the probability of his death down to 1/3. A similar argument holds from the soldier's standpoint. By using his optimal strategy he keeps the payoff at 1/3 and cannot hope to do better because the gunner has a way of making it at least 1/3. As a further exercise, readers can try to prove there are no optimal strategies other than those explained here.

"The process of surmising the solution is not as hard as it looks," Isaacs adds. "The reader can so convince himself by generalizing this solution to the same game but with n foxholes. For odd n the preceding solution carries over in an almost obvious way, but with even n one encounters some modest novelty."

CHAPTER 4

Factorial Oddities

MATHEMATICAL FORMULAS, particularly combinatorial formulas, are sometimes sprinkled with exclamation marks. These are not expressions of surprise. They are operational symbols called factorial signs. A factorial sign follows a whole number or an expression for such a number, and all it means is that the number is to be multiplied by all smaller whole numbers. For example, $4!$, read as "factorial four," is the product of $4 \times 3 \times 2 \times 1$. (In older books factorial n is symbolized $\underline{n|}$.)

Why are factorials so important in combinatorial mathematics and in probability theory, which relies so heavily on combinatorial formulas? The answer is simple: Factorial n is the number of different ways that n things can be arranged in a line. Visualize four chairs in a row. In how many different ways can four people seat themselves? There are four ways to fill the first chair. In each of those ways there are three ways the second chair can be occupied, and so there are 4×3, or 12, ways to fill the first two chairs. For each of those ways there are two ways to occupy the third chair, and so there are $4 \times 3 \times 2$, or 24, ways to fill the first three chairs. In each of those 24 instances there is only one person left to take the fourth chair. The total number of distinct ways of occupying the four chairs is therefore 4!, or $4 \times 3 \times 2 \times 1 = 24$.

The same reasoning shows that 52 playing cards can be made into a deck in 52! different ways, a number of 68 digits that begins 806581 . . .

What is the probability that a bridge player will be dealt 13 spades? We first determine the number of different bridge hands. Since the order in which the cards are dealt is irrelevant, we want to know not the permutations but the number of different *combinations* of 13 cards that can be made with the 52 cards of the deck. The formula for a combination of n elements, taken r at a time, is $n!/r!(n-r)!$. With n equaling 52 and r equaling 13, we have $52!/(13! \times 39!)$, which works out to 635,013,559,600. A bridge player can therefore expect to be dealt all spades once in every 635,013,559,600 deals. His chance of getting 13 cards of the same suit, not necessarily spades, is four times this, or once in every 158,753,389,900 hands. The chance that such a hand will be dealt to one of the four players is 1 over a number a trifle less than one-fourth of the previous number. (It is not exactly one-fourth because we have to consider the possibility that two or more players may each have a one-suit hand.) It works out to what is still such a stupendously low probability, 1/39,688,347,497, that one is led to conclude that reports of such hands, which turn up in some newspaper around the world about once a year, are almost certainly false reports, hoaxes, or the result of a dealer accidentally giving a new deck two perfect riffle shuffles. (A cut would not disturb the cyclic ordering.) It is a curious fact (as Norman T. Gridgeman, a Canadian statistician, has pointed out) that although there are frequent reports of four perfect hands, there are *no* reports of two perfect hands, which is millions of times more probable.

It is easy to see from these elementary examples that it is much simpler to express large factorials by using the factorial sign than to write out the entire number. Indeed, factorials increase in size at such a rapid rate [*see Figure 8*] that until the advent of high-speed computers exact values for factorials were known only up to about 300!, except for a dozen or so higher factorials which someone bothered to compute.

FIGURE 8
Factorials from 0 to 20

0! = 1
1! = 1
2! = 2
3! = 6
4! = 24
5! = 120
6! = 720
7! = 5,040
8! = 40,320
9! = 362,880
10! = 3,628,800
11! = 39,916,800
12! = 479,001,600
13! = 6,227,020,800
14! = 87,178,291,200
15! = 1,307,674,368,000
16! = 20,922,789,888,000
17! = 355,687,428,096,000
18! = 6,402,373,705,728,000
19! = 121,645,100,408,832,000
20! = 2,432,902,008,176,640,000

Note on the chart that 7! = 5,040, a very interesting number. In Book 5 of his *Laws*, Plato gives this as the population of an ideal city. His argument is that 5,040 has an unusually large number of factors (59, including 1 but not 5,040), which makes for efficient division of the populace for purposes of taxes, land distribution, war, and so on. (Plato was probably not aware that 7,560 and 9,240 each have 63 proper divisors, the maximum possible for a number of four or fewer digits. For this and more on Plato's number, see my book *The Magic Numbers of Doctor Matrix* [Prometheus, 1985], Chapter 14.)

Higher factorials can be approximated by Stirling's formula, named after James Stirling, an eighteenth-century Scottish mathematician:

$$n! \cong n^n e^{-n} \sqrt{2\pi n}.$$

It is a strange formula, involving the two best-known transcendental numbers, pi and e. The formula's "absolute error" (the difference between a factorial's true value and its approximation) increases as factorials get larger, but the "percentage of error" (absolute error divided by the true value) steadily decreases.

For practical purposes the formula gives excellent approximations of high factorials, but mathematicians have a compulsion to know things precisely. Just as mountain climbers will scale a peak or space explorers go to the moon "because it is there," so mathematicians with access to computers have irrepressible urges to explore the "outer space" of enormous numbers. (Precisely in what sense we can say a large factorial, never before computed, "is there" is a question that mathematicians answer differently, depending on their philosophy of mathematics.) It was the computer that made it possible, by monitoring and interpreting signals, to take close-up looks at the surface of the moon and Mars. Those same computers have also made it possible to take close-up looks at large factorials, gigantic numbers that have been known for centuries only in a vague, out-of-focus way.

The remarks in the previous paragraph are taken from a communication received from Robert E. Smith, director of computer applications at the Control Data Institute in Minneapolis. He writes that he was exploring the outer fringes of factorials when it occurred to him that it would be pleasant to print on his Christmas card, in the form of a Christmas tree, one of his mammoth factorial numbers. It would be necessary, of course, to have the computer print one digit for the top of the tree, then a row of three digits, then a row of five, and so on. Are there factorials with the proper number of digits so that such a printout would form a perfect tree? Yes, there is an infinity of them. The table [*see Figure 8*] shows, for example, that 12! has nine digits. It can be printed in tree form like this:

4
790
01600

A tree factorial obviously must have a number of digits that is a partial sum of the infinite series $1 + 3 + 5 + 7 + \ldots$ A glance at the square array of spots below shows that all such sums are perfect squares:

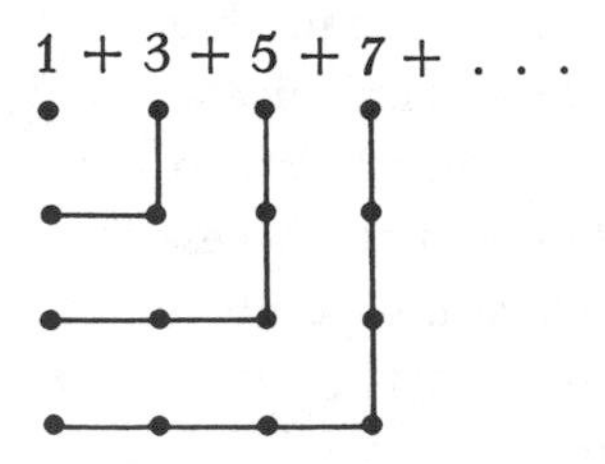

Smith's task, therefore, was to program a computer to search for large factorials that have a square number of digits but instead of printing them in square formation to print them in lines of 1, 3, 5, 7 . . . digits to form a tree. This was done. The computer tested all factorials up to approximately 1,000! (and a few larger ones) and found exactly 20 factorials less than 1,000! that contained a square number of digits [*see Figure 9*].

In the printout of 105! from Smith's computer, note the bottom row of zeros [*see Figure 10*]. If the reader will study the series $1 \times 2 \times 3 \times 4 \times \ldots$, he will see that every multiplication by a number ending in 5 will add one or more zeros to the running product, and that every multiplier ending in one or more zeros will add additional zeros. Because these terminal zeros cannot be lost by later multiplications they are cumulative, piling up steadily at the ends of factorials as they get larger. Factorial 105 has a "tail" of 25 zeros. Stirling's formula,

Factorials	Number of Digits
7	4
12	9
18	16
32	36
59	81
81	121
105	169
132	225
228	441
265	529
284	576
304	625
367	784
389	841
435	961
483	1,089
508	1,156
697	1,681
726	1,764
944	2,401

FIGURE 9

Tree factorials less than 1,000!

```
            1
           081
          39675
         8240290
        900504101
       30580032964
      9720646107774
     902579144176636
    57322653190990515
   3326984536526808240
  339776398934872029657
 99387290781343681609728
0000000000000000000000000
```

FIGURE 10

Tree printout of the 169 digits of 105!

applied to 105!, gives an approximation that is roughly equal to 1,081, followed by 165 zeros. When this blurry result is compared with the exact result shown in the illustration, one sees that, in Smith's words, "using Stirling's formula for arriving at large factorials is analogous to a blind man trying to visualize an elephant by grasping a couple of inches of its trunk in one hand and the tip of its tail in the other.

"It may surprise the reader to know," Smith continues, "that computers cannot calculate results this large in one piece. That is, the capacities of individual computer cells are soon exceeded, so normal computer arithmetic cannot be used." The trick, he explains, is to use several internal "bins," each of which is allowed to hold t digits of a result. "After each multiplication is performed, excess digits beyond the t digits, in each bin, are removed and added to the bin on their immediate left. Then the digits in all the bins, which form the factorial result, are printed." A fuller explanation of this program will be found in Smith's book *The Bases of FORTRAN*, published in 1967 by the Control Data Institute. Another large tree factorial computed by this method is shown in Figure 11.

When the number of digits in a factorial is equal to two consecutive squares, such as 35!, which has $4^2 + 5^2 = 41$ digits, it can be printed as a diamond. One simply turns the smaller tree of 4^2 digits upside down and fits it against the base of the 5^2 tree like this:

```
    1
   033
  31479
 6638614
4929 6665
 1337523
  20000
   000
    0
```

I have blanked out the center digit to give the reader a small problem. What is the missing digit? By applying a simple trick known to most accountants, and which every reader of this book ought to know, the missing digit can be found quickly without a single multiplication.

```
                                 5
                                119
                               90692
                              7755879
                             266003615
                            25819185379
                           7984360677298
                          470133958906714
                         46011174633964398
                        5839112233165772956
                       548496166254935516795
                      14565079522588677608012
                     6423489045662147453126349
                    825790036437158643266482002
                   88113505694916924243929121639
                  7995123320680205388149829536720
                 697546589338105120020005674705145
                28641409978978956631664608452253922
               2182139322091260889711710217500934598
              659546487929459214735007200769105667735
             54074289548655659977226200540160335058131
            8365384235510714071491098835812736588922795
           511456461421254773804907853073384484888784090
          75030962875912509521999525292598359880846423952
         3931204111818280979213544777644751538435208774603
        088477116032223651164439419220002073567325180151958
       35354728897604905269289015307797618984464654042934912
      7882733479825616955531216107050271401259459875249508169
     440013327395316887000833911764483284987619075088343797786
    47371945157918046252226969546616811434035461815792968273198
   2545625613705049834238544557702694536385292145346080336071424
  289160111720849018903249047529128422886467764267877861568498090
 42964480000000000000000000000000000000000000000000000000000000000
0000000000000000000000000000000000000000000000000000000000000000000
```

FIGURE 11

Tree printout of the 1,156 digits of 508!

There are many other basic geometric shapes in which certain factorials can be printed. Smith's computer printed 477! as a hexagon with 17 digits on each side [*see Figure 12*]. Finally, to stagger the reader with one of Smith's computer's supermonstrosities, there is 2,206! in octagonal form [*see Figure 13*]. If someone had predicted fifty years ago that before the century ended this factorial would be written out in full, digit by digit, most mathematicians would have laughed at so preposterous a prophecy.

```
                17108972589718074
               1439528307936299026
              080765545554532458183
             43255130543516432376912
            4663791911119657860822050
           367340495642348613717749611
          38104459104482535212494659899
         5225079402598873366451131040234
        240130493689852679573590918519290
       66647636392705738600295487428650940
      0535103538524596394743595531728001643
     083783948745781956212836911156587085000
    40781396853030778257813849856692950471963
   5089328018573725755534194119396813233357487
  709737509271413007324171020350516977549843435
 61187933295519151457453789138048055187827977590
7750007855795139817496078270462761613125177420579
 97170554688538703689036095806399241086011592997
  020790226888203087101533653915806041722653430
   0377642434651424325601245917031000886439794
    86942002854170097571338930915447098888372
     333024657251637441276280296188483408232
      2723195014038951851520634322622612616
       12431271509190879459978732133255390
        601413833379281814639023615443036
         2338368861798856260050552801204
          22598617062894203619648023806
           809643832836096000000000000
            0000000000000000000000000
             00000000000000000000000
              000000000000000000000
               0000000000000000000
                00000000000000000
```

FIGURE 12

Hexagon printout of 477! (1,073 digits)

YOU ARE LOOKING AT FACTORIAL 2206 THERE ARE 6421 DIGITS IN THIS ANSWER

```
1025729611602273508352741640301
16267940732286960838307447113874 6
30490176947592788989158028363440992
4881004274871081412845389545578617964
937451453403559961733083620099625538065
41010396598345035653688604954874028604784
6935414801400008952354538805490908409959861
660246045191650260507136760688118903350996713
21823947514423961017742558259080494678177458085
2586022822207646670820883488582338159458352125708
362194967520635351099568538807160144980504806697217
60361821401601138846975754268977469634297805191393877
0721740965493361497666505546284509191953032898888959567
069201586371013497509110063361133485051253334853981163412
07851202228866097513788527648919064612804715051516533550808
5628864147138448743521002113548909441417847473055608205047228
050888506002351831385707706597375486196289582204458204639226419
39988685801691888740823242074655630309966434682847648783012369690
0949679573713722032502963934989938775237593102275000626771342793017
571237823459954555639364520012107232014386673952152555563687059868597
30561699128318275778511690671009105944054429948225877769250180772939089
5482468231085915745358203314543675656839465279720697501109412886408456325
452410148100374780426054880997809853185180536108897611555630604942279452179
84296686240652468994135036474242868453665349623925844722715297861094652833533
6243076317214023025703473585182809668132851399986647037941978752777468292847695
603640621614001511781845118088875779459341557554605484271581941683408763840520361
23170898026795922847394635627557990754271854045696455075681612990097706515709337585
3609457155707951999956996602370489306141334822832959737146934986498711106505143623006
463954096916259897686557876127375781093939500600724947477213915033139433655578405917439
66529210437027529365792843517778279323319868406572635522503159691924609965063914289434510
7019821565442168516926969752389318995661246814055912699808919159203294055178987900314759208
0442506495067999489407758571492797940488049296440489928011632417128028690854571931206525088
0641218070711371216774493418451658654662421419901954748914071330317835008057673531461465561
2768515633920521576618161998879098027906520489507115754887782147626888335639598691326086662
8591561763561263074922316182055408616512641833590647698679206429535534489790054789746645046
7541595913406494278781641172660577772496746978451472658691427962204322571171000546469639193
2790901159275847484128114857934498847063422270277559602059269990527077182952388889333136592
5112259111214461121960558583085300900857728673001457908605480987931391634861299625358963146
9354877285883646562527594000195845750815246146536989440348537192331062397234558495885351131
0486319190754485743708055141816253285423820670998154301257121647181580736841247433302909291
8978304729883750528094105685468933501043414202907313725590875531859893702065889477849388404
1409345973933364835255409965082222341074607394942553036358497877027519189446464290292514742
8894719069197344614264735325576405364746890544013692524832899026218553163092629405073307432
5883237879257553223715273633812159745601142017086546047693697449507986197545225600849607508
4225614110584212738361884639106673278902989788537027787173903234357897043773626379633919708
2650317682667625547582587775747027139684279901717692515347217940718330972281022406020576955
1923109334083392446906107789059722221024694681063461315307372556082106306372694424873419069
0964340389530663549733052473631156529648322803073354434281448023017979554302704018756348368
8235987859335034065876692541582476375532676240789963230134591939936849075772844192830282673
5947374533149522043979463889533209159979640809623687593870067491951481812767444483279372007
2479979995918726706098991265562174581709978283817056885289631559290783185857144934100631908
9346302000944243304245025830533367966786168167762945518348245854102054733012367306021265699
7144412744968191763657240549320164830818358694058022230045712574068580666063952523933901667
7230224963193557961156169360046766935581419151926024167053911783185412817174305853343800679
1914069190803158006986526614048105699422168264057749481333683845736260715525170992590292274
4512642655765465531121253764415693728736415931091968744923431544233653466036812168293962954
9767299210770870181089859892303670680851719783785348200229620448896647181023724949915937363
8125724983730036346858277218777076810719393224395941146613075583124238948926153797306851203
3222908555754737090939960701071272925206188572711353732688814416502259766751986220308762254
0147673137413584101747534961199349889138751673014525041430085687942534349828532365955990586
3059290311067608656677397188828409975102079430788883571949249071704661577344739732272034246
37104435122049746259745064632630405327181489715562072507013420213784243184193841952049019
389152859513288157006180759969996691548238462575683756080195840926147215172703045410726
1571221590491814008464517752380714664838663797936798266141064998973891352888615965540
98255595359708203960431881516899044902069946858812271164189344778402447747132779673
726734254206089602373228992235406035343758920286285382782483616919760583195041901
0786025652398432473377918786067668522964469358193816182564719257636482538175360 4
03238462202327486151772182306892159147076520062053518885595951392766191659916
67545893045431742688059944064925383518255904913468078411329134916930248201 7
38031651030450833092417520278619904144657169102128566900483402128010665 66
71488714187828459473979416544321600570091801616015052259641365801993579
244504186187559237535573796932130060786723802037406586217527166133087
0930181945058128239446205417863686242419793225560961759772203435083
0138135915753136145678366144068394600394572699435664121511329101 6
15863145785714176519103070180866052902762823446442960463008901 4
8953434120066141769048632529168891605980149568222227367224697
1334697124770615322089499322046034601267595656592498513778 7
34801684974735804507040078862939942473633835061096833117 0
076845670400000000000000000000000000000000000000000000
000000000000000000000000000000000000000000000000000
0000000000000000000000000000000000000000000000000
00000000000000000000000000000000000000000000000
000000000000000000000000000000000000000000000
0000000000000000000000000000000000000000000
00000000000000000000000000000000000000000
000000000000000000000000000000000000000
0000000000000000000000000000000000000
00000000000000000000000000000000000
000000000000000000000000000000000
0000000000000000000000000000000
00000000000000000000000000000
000000000000000000000000000
```

FIGURE 13
Octagon printout of 2,206! (6,421 digits!)

Factorials, as one would expect, are closely related to primes. The most famous of the many elegant formulas that link the two kinds of number is known as Wilson's theorem, after an eighteenth-century English judge, Sir John Wilson, who hit on the formula when he was a student at Cambridge. (It later developed that Leibniz knew the formula.) Wilson's theorem says that $(n-1)!+1$ is divisible by n if, and only if, n is a prime. For example, if n equals 13, then $(n-1)!+1$ becomes $12!+1=479{,}001{,}601$. It is easy to see that 12! is not a multiple of 13, because 13 is prime and the factors of 12! do not include 13 or any of its multiples. But, astonishingly, the mere addition of 1 creates a number that *is* divisible by 13. Wilson's theorem is one of the most beautiful and important theorems in the history of number theory, even though it is not an efficient way to test primality.

There are many simply expressed but difficult problems about factorials that have never been solved. No one knows, for example, if a finite or an infinite number of factorials become primes by the addition of 1, or even how many become squares by the addition of 1. (We are concerned now with the number itself, not its number of digits.) It was conjectured back in 1876 by H. Brocard, a French mathematician, that only three factorials—4!, 5!, and 7!—become squares when they are increased by 1. Albert H. Beiler, in *Recreations in the Theory of Numbers*, says that this has been investigated by computers up to factorial 1,020 without finding any other solution, but Brocard's conjecture remains unproved.

It is easy to find factorials that are the products of factorials but hard to find them if the factorials to be multiplied are in arithmetic progression and still harder to find them if they are consecutive. Only four consecutive instances are known: $0!\times 1! = 1!$, $0!\times 1!\times 2! = 2!$, $1!\times 2! = 2!$, and $6!\times 7! = 10!$.

It should be explained, in connection with the first two solutions, that 0! is defined as 1 in spite of the fact that 1 is also the value of 1!. Strictly speaking, 0! is meaningless, but by making it 1 many important formulas are kept free of anomalies. The basic identity $n! = n(n-1)!$, which means that factorial n equals n times the factorial of the preceding integer, is true for all positive values of n only if $0! = 1$. Consider the formula given earlier, $n!/r!(n-r)!$, to determine how many combinations there are of two things taken two at a time. The answer, of course, is 1. The formula gives that answer only if 0! equals 1. If 0! equals 0, the formula becomes meaningless because it simplifies to a division by zero. The famous binomial theorem, discovered by Isaac Newton, is another classic instance of a basic formula that can be debugged, so to speak, by defining factorial 0 as 1.

The quaint problem of finding whole numbers equal to the sum of the factorials of their digits has recently been solved. There are four solutions. Two are trivial: $1 = 1!$ and $2 = 2!$. The largest example was found in 1964 by Leigh Janes of Houston, using a computer: $40{,}585 = 4! + 0! + 5! + 8! + 5!$. Can the reader find the remaining solution? It can be expressed as $A! + B! + C! = ABC$. Each letter is a different digit.

Of many classic recreational problems for which factorials provide elegant solutions, I select one from graph theory. A man who lives at the top left corner of a rectangular grid of city blocks [*see Figure 14*] works in an office building at the bottom

FIGURE 14
Route problem involving factorials

right corner. It is clear that the shortest path along which he can walk to work is 10 blocks long. Bored with following the same route each day, he begins to vary it. How many different 10-block routes are there connecting the two spots? And what is a compact formula for the number of different minimum routes joining the diagonally opposite corners of any rectangular area of blocks? (Here is a hint: The number of different arrangements, or permutations, of n objects of which a objects are identical and the remaining b objects are also identical is $n!/a!b!$.)

ADDENDUM

THE FORMULA given at the close of the chapter is a special case of an important general formula. The number of different arrangements of n objects, of which a are identical, b are identical, c are identical, and so on, is

$$\frac{n!}{a!b!c! \ldots}$$

The formula is easily explained. The number of permutations of n objects, all different, is $n!$. If a objects are alike, it makes no difference how *they* are permuted, so we divide by $a!$ to eliminate all the permutations that are identical because of the identities of the a objects. Similarly, we also divide by $b!$ to eliminate all permutations that are identical because of the identities of the b objects, and so on for permutations of other identical objects. The letters of *MISSISSIPPI*, for example, can be permuted 11! ways if each letter is considered distinct. But if we treat the four *I*'s as identical, as well as the four *S*'s and two *P*'s, the number of permutations is $11!/(4! \times 4! \times 2!) =$ 34,650.

Many readers called my attention to an oddity involving factorials and the calculus of finite differences. Start with the sequence of consecutive numbers raised to the power of n, where n is any non-negative integer:

$$1^n \ 2^n \ 3^n \ 4^n \ 5^n \ 6^n$$

The nth row of differences is an endless repetition of $n!$. Note that this holds even when $n = 0$.

Donald E. Boynton, James Cassels, and Rear Admiral Robert S. Hatcher each sent the following simple method of calculating the number of zeros in the tail of any factorial. The method seems not to be well known. If a multiplier is divisible by 5 but not by 5^2 it adds one zero to the product. If divisible by 5^2 but not by 5^3 it adds two zeros, if divisible by 5^3 but not 5^4 it adds three zeros, and so on. Therefore, the number of terminal zeros of $n!$ can be found by dividing n by 5, discarding the remainder, dividing the quotient by 5, discarding the remainder, and repeating this process until the quotient is less than 5. The sum of all the quotients is the number of zeros. Example: 2,206!, as shown in Figure 13, has 549 zeros at the end. The successive quotients, after 2,206 has been repeatedly divided by 5 (remainders discarded), are 441, 88, 17, and 3. They sum to 549.

The problem of determining the last nonzero digit of a large factorial calls for a more complicated algorithm. I leave this for the interested reader to work out. What, for instance, is the last nonzero digit of 1,000!?

ANSWERS

THE MISSING CENTER DIGIT in the diamond-shaped factorial is easily found by recalling that every multiple of 9 has a digital root of 9; if one keeps summing the digits of the number, casting out 9's as he goes along, the final digit must be 9. Every factorial higher than 5! is a multiple of 9 because 6! has 3 and 6 as factors and 3 times 6 is 18, a multiple of 9. Therefore to find a missing digit in any factorial greater than 5! one simply obtains the digital root of the mutilated factorial and subtracts it from 9 to get the missing digit. If the mutilated factorial has a digital root of 9, the missing digit could be either 0 or 9, but in this case the digital root of the mutilated diamond factorial is 3, so there is no ambiguity. The missing center digit must be 6.

Had there been ambiguity, the missing digit is easily found by using a familiar test for divisibility by 11. All factorials greater than 10! are obviously exact multiples of 11. If a number is a multiple of 11, the sum of its digits in even places either equals the sum of its digits in odd places, or the sums differ by a multiple of 11. This test leaves no ambiguity about any missing digit in a factorial greater than 10!

$A! + B! + C! = ABC$ has the unique solution $1! + 4! + 5! = 145$. For a proof that 1, 2, 145, and 40,585 are the only positive integers, each equal to the sum of the factorials of its digits, see George D. Poole, "Integers and the Sum of the Factorials of Their Digits," *Mathematics Magazine*, Vol. 44, November 1971, pages 278–79.

To find the number of different minimum-length routes from one corner of a rectangular section of city blocks to the diagonally opposite corner, consider that if the rectangular section is a blocks long and b blocks wide, the minimum path that connects diagonally opposite corners is a plus b. Call this sum n. Every n-length path from corner to corner can now be expressed as a chain of n symbols, of which a symbols will be identical (indicating a block's travel lengthwise toward the goal) and the remaining b symbols will be identical (indicating a block's travel widthwise to the goal). If we let a penny stand for a movement of one block lengthwise and a dime for a movement of one block widthwise, then the number of different routes will equal the number of different ways that a pennies and b dimes can be arranged in a row. Every distinct route can be put in one-to-one correspondence with a permutation of the n coins, and every permutation of the coins corresponds to one of the routes.

The hint was the formula $n!/a!b!$ for the number of ways of arranging n objects in a row, of which a are identical and the remaining b are identical. The rectangle is six blocks long and four blocks wide, therefore the problem of determining the number of different routes is isomorphic with the problem of finding the number of different ways six pennies and four dimes can be placed in a row. The answer is $10!/(6! \times 4!) = 210$.

The problem ties in with the discussion of Pascal's triangle in Chapter 15 of my *Mathematical Carnival* (Mathematical Association of America, 1989), as the reader will discover if he labels each intersection with the number of different minimum-length routes from the upper left corner to that intersection. The answer, 210, is found on the triangle simply by starting at the top of the triangle, moving six (or four) steps down one side, then turning and going down four (or six) steps in the other diagonal direction.

CHAPTER 5

The Cocktail Cherry and Other Problems

1. THE COCKTAIL CHERRY

THIS IS one of those rare, delightful puzzles that can be solved at once if you approach it right, but that is subtly designed to misdirect your thoughts toward the wrong experimental patterns. Intelligent people have been known to struggle with it

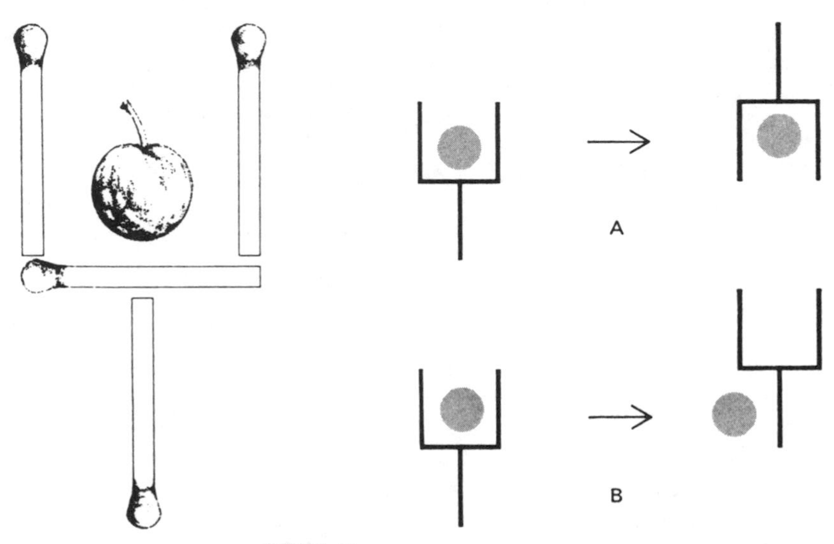

FIGURE 15
The puzzling Manhattan

for twenty minutes before finally deciding that there is no solution.

Place four paper matches on top of the four matches that form the cocktail glass in Figure 15. The problem is to move two matches, and only two, to new positions so that the glass is re-formed in a different position and the cherry is *outside* the glass. The orientation of the glass may be altered but the empty glass must be congruent with the one illustrated. The drawing at *A* shows how two matches can be moved to turn the glass upside down. This fails to solve the problem, however, because the cherry remains inside. The drawing at *B* shows a way to empty the glass, but this does not solve the puzzle either because three, rather than two, matches have been moved.

2. THE PAPERED CUBE

WHAT IS the largest cube that can be completely covered on all six sides by folding around it a pattern cut from a square sheet of paper with a side of three inches? (The pattern must, of course, be all in one piece.)

3. LUNCH AT THE TL CLUB

EVERY MEMBER of the TL Club is either a truther, who always tells the truth when asked a question, or a liar, who always answers with a lie. When I visited the club for the first time, I found its members, all men, seated around a large circular table, having lunch. There was no way to distinguish truthers from liars by their appearance, and so I asked each man in turn which he was. This proved unenlightening. Each man naturally assured me he was a truther. I tried again, this time asking each man whether his neighbor on the left was a truther or a liar. To my surprise each told me the man on his left was a liar.

Later in the day, back home and typing up my notes on the luncheon, I discovered I had forgotten to record the number of men at the table. I telephoned the club's president. He told me the number was 37. After hanging up I realized that I could not be sure of this figure because I did not know whether the president was a truther or a liar. I then telephoned the club's secretary.

"No, no," the secretary said. "Our president, unfortunately, is an unmitigated liar. There were actually 40 men at the table."

Which man, if either, should I believe? Suddenly I saw a simple way to resolve the matter. Can the reader, on the basis of the information given here, determine how many men were seated at the table? The problem is derived from a suggestion by Werner Joho, a physicist in Zurich.

4. A FAIR DIVISION

TWO BROTHERS inherited a herd of sheep. They sold all of them, receiving for each sheep the same number of dollars as there were sheep in the herd. The money was given to them in $10 bills except for an excess amount, less than $10, that was in silver dollars. They divided the bills between them by placing them on a table and alternately taking a bill until there were none left.

"It isn't fair," complained the younger brother. "You drew first and you also took the last bill, so you got $10 more than I did."

To even things up partially the older brother gave the younger one all the silver dollars, but the younger brother was still not satisfied. "You gave me less than $10," he argued. "You still owe me some money."

"True," said the older brother. "Suppose I write you a check that will make the total amounts we each end up with exactly the same."

This he did. What was the value of the check? The information seems inadequate, but nevertheless the question can be answered.

Ronald A. Wohl, a chemist at Rutgers University, called my attention recently to this beautiful problem, which he had found in a French book. Later I discovered in my files a letter from

Carl J. Coe, a retired mathematician at the University of Michigan, discussing essentially the same problem, which he said had been making the rounds among his colleagues in the 1950's. I suspect it is still not widely known.

5. TRI-HEX

TICKTACTOE is played on a pattern that can be regarded as nine cells arranged in eight rows of three cells to a row. It is possible, however, to arrange nine cells in nine or even ten rows of three. Thomas H. O'Beirne of Glasgow, author of *Puzzles and Paradoxes* (Oxford, 1965), experimented with topologically distinct patterns of nine rows to see if any were suitable for ticktacktoe play. He found trivial wins for the first player on all regular configurations except the one shown in Figure 16.

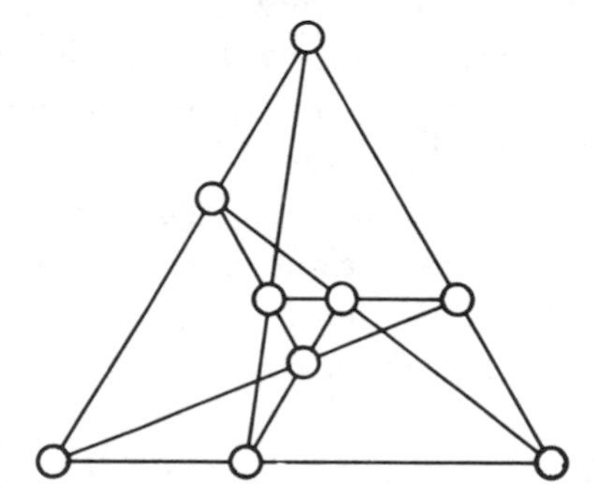

FIGURE 16
The game of Tri-Hex

To play Tri-Hex, as O'Beirne calls this game, one player can use four pennies and the other four dimes. No fifth move is allowed the first player. Players take turns placing a coin on a spot, and the first to get three of his coins in a row wins. If both players make their best moves, is the game a win for the first player or the second, or is it a draw as in ticktacktoe?

The role played by such configurations as this in modern geometry is entertainingly discussed by Harold L. Dorwart in *The Geometry of Incidence* (Prentice-Hall, 1966) and in the instruction booklet for his puzzle kit, *Configurations*, now available from the makers of the logic game WFF'N PROOF. In addition to its topological and combinatorial properties, the pattern shown here has an unusual metric structure: every line of three is divided by its middle spot into segments with lengths in the golden ratio.

6. LANGFORD'S PROBLEM

MANY YEARS AGO C. Dudley Langford, a Scottish mathematician, was watching his little boy play with colored blocks. There were two blocks of each color, and the child had piled six of them in a column in such a way that one block was between the red pair, two blocks were between the blue pair, and three were between the yellow pair. Substitute digits 1, 2, 3 for the colors and the sequence can be represented as 312132.

This is the unique answer (not counting its reversal as being different) to the problem of arranging the six digits so that there is one digit between the 1's and there are two digits between the 2's and three digits between the 3's.

Langford tried the same task with four pairs of differently colored blocks and found that it too had a unique solution. Can the reader discover it? A convenient way to work on this easy problem is with eight playing cards: two aces, two deuces, two threes, and two fours. The object then is to place them in a row so that one card separates the aces, two cards separate the deuces, and so on.

There are no solutions to "Langford's problem," as it is now called, with five or six pairs of cards. There are 26 distinct solutions with seven pairs. No one knows how to determine the number of distinct solutions for a given number of pairs except by exhaustive trial-and-error methods, but perhaps the reader can discover a simple method of determining if there *is* a solution.

7. OVERLAP SQUARES

IN 1950, when Charles W. Trigg, dean emeritus of Los Angeles City College, was editing the problems department of *Mathematics Magazine*, he introduced a popular section headed

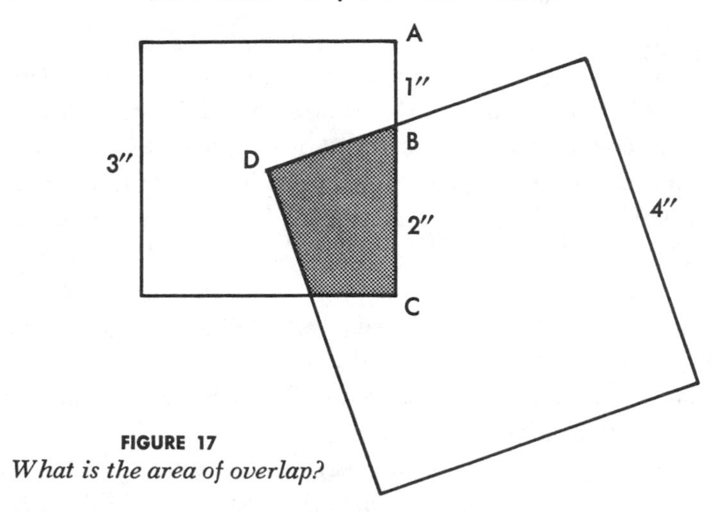

FIGURE 17
What is the area of overlap?

"Quickies." A quickie, Trigg then explained, is a problem "which may be solved by laborious methods, but which with proper insight may be disposed of with dispatch." His 1967 book, *Mathematical Quickies* (reprinted by Dover in 1985), is a splendid collection of 270 of the best quickies he encountered or invented in his distinguished career as a problem expert.

In one quickie from Trigg's book [*see Figure 17*] the smaller square has a side of three inches and the larger square a side of four. Corner *D* is at the center of the small square. The large square is rotated around *D* until the intersection of two sides at point *B* exactly trisects *AC*. How quickly can you compute the area of overlap (shown shaded) of the two squares?

8. FAMILIES IN FERTILIA

People who have three daughters try once more
And then it's fifty-fifty they'll have four.
Those with a son or sons will let things be.
Hence all these surplus women. Q.E.D.

This "Note for the Scientist," by Justin Richardson, is from a Penguin collection called *Yet More Comic & Curious Verse*, selected by J. M. Cohen. Is the expressed thought sound?

No, although it is a commonly encountered type of statistical fallacy. George Gamow and Marvin Stern, in their book *Puzzle-Math* (Viking, 1958), tell of a sultan who tried to increase the number of women available for harems in his country by passing a law forbidding every mother to have another child after she gave birth to her first son; as long as her children were girls she would be permitted to continue childbearing. "Under this new law," the sultan explained, "you will see women having families such as four girls and one boy; ten girls and one boy; perhaps a solitary boy, and so on. This should obviously increase the ratio of women to men."

As Gamow and Stern make clear, it does nothing of the sort. Consider all the mothers who have had only one child. Half of their children will be boys, half girls. Mothers of the girls will then have a second child. Again there will be an even distribution of boys and girls. Half of those mothers will go on to have a third child and again there will be as many boys as there are girls. Regardless of the number of rounds and the size of the families, the sex ratio obviously will always be one to one.

Which brings us to a statistical problem posed by Richard G. Gould of Washington. Assume that the sultan's law is in effect and that parents in Fertilia are sufficiently potent and long-lived so that every family continues to have children until there is a son, and then stops. At each birth the probability of a boy is one-half. In the long run, what is the average size of a family in Fertilia?

9. CHRISTMAS AND HALLOWEEN

PROVE (asks Solomon W. Golomb) that Oct. 31 = Dec. 25.

10. KNOT THE ROPE

OBTAIN a piece of clothesline rope about five and a half feet long. Knot both ends to form a loop as shown in Figure 18. Each loop should be just large enough to allow you to squeeze a hand through it. With the loops around each wrist and the rope stretched between them, is it possible to tie a single overhand knot in the center of the rope? You may manipulate the rope any way you like, but of course you must not slide a loop off your wrist, cut the rope, or tamper with either of the existing knots. The trick is not well known except to magicians.

FIGURE 18
Rope for the knot problem

ANSWERS

1. FIGURE 19 shows how two matches are moved to re-form the cocktail glass with the cherry outside.

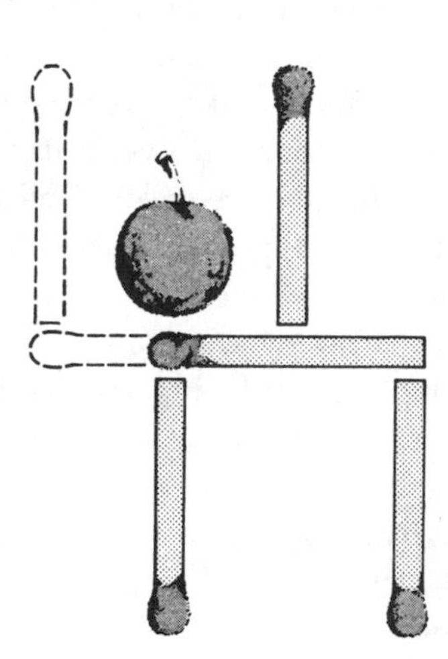

FIGURE 19
Solution to match problem

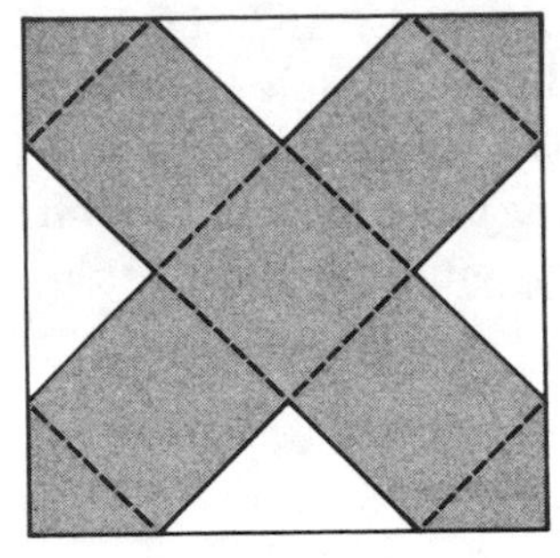

FIGURE 20
Nonoverlap solution

2. If overlapping of paper is not allowed, the largest cube that can be folded from a pattern cut from a square sheet of paper with a three-inch side is a cube with a side that is three-fourths of the square root of 2. The pattern, shown in Figure 20, is folded along dotted lines.

In stating the problem, however, I did not forbid overlapping. This was not an oversight. It simply did not occur to me that overlapping would permit better solutions than the one above, which was given as the problem's only answer. John H. Halton, a mathematician at the University of Wisconsin, was the first to send a cut-and-fold technique by which one can approach as closely as one wishes to the ultimate-size cube with a surface area equal to the area of the square sheet! (Three readers, David Elwell, James F. Scudder, and Siegfried Spira, each came close to such a discovery by finding ways to cover a cube larger than the one given in the answer, and George D. Parker found a complete solution that was essentially the same as Halton's.)

Halton's technique involves cutting the square so that opposite sides of the cube are covered with solid squares and the rest of the cube is wrapped with a ribbon folded into one straight strip so that the amount of overlap can be as small as one wishes. In wrapping, the overlap can again be made as small as one pleases simply by reducing the width of the ribbon. Assuming infinite patience and paper of infinitely small molecular dimensions, as Halton put it, this procedure will cover a cube approaching as closely as desired to the limit, a cube of side $\sqrt{3}/\sqrt{2}$.

Fitch Cheney, a mathematician at the University of Hartford, found another way to do the same thing by extending the

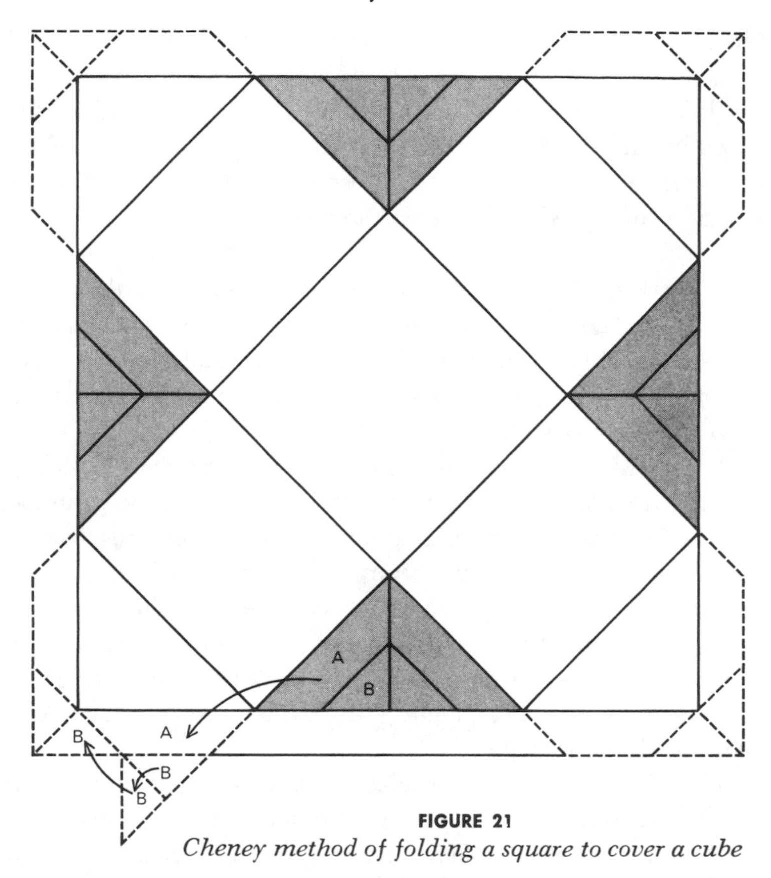

FIGURE 21
Cheney method of folding a square to cover a cube

pattern in Figure 20. By enlarging the center square as shown in Figure 21, he makes the four surrounding squares rectangles and the four corner triangles correspondingly larger. The shaded areas, cut as shown, can be folded as indicated to extend each corner triangle. (A turns over once, B three times.) The result: a pattern exactly like the one given, except that it is larger. Since the unavoidable overlap can be made as small as one wants, it is clear that this method also approaches the $\sqrt{3}/\sqrt{2}$ cube as a limit.

3. If each man at a circular table is either a truther or a liar, and each says that the man on his left is a liar, there must be an even number at the table, arranged so that truthers and liars al-

ternate. (No arrangement of an odd number of truthers and liars is possible without at least one man describing the man on his left as a truther.) Consequently the club's president lied when he said the number was 37. Since the secretary called the president a liar, he must have been a truther. Therefore he spoke truthfully when he gave the number as 40.

4. We were told that the two brothers who inherited a herd of sheep sold each sheep for the same number of dollars as there were sheep. If the number of sheep is n, the total number of dollars received is n^2. This was paid in $10 bills plus an excess, less than 10, in silver dollars.

By alternately taking bills, the older brother drew both first and last, and so the total amount must contain an odd number of 10's. Since the square of any multiple of 10 contains an even number of 10's, we conclude that n (the number of sheep) must end in a digit the square of which contains an odd number of 10's. Only two digits, 4 and 6, have such squares: 16 and 36. Both squares end in 6, and so n^2 (the total amount received for the sheep) is a number ending in 6. The excess amount consisted of six silver dollars.

After the younger brother took the $6 he still had $4 less than his brother, so to even things up the older brother wrote a check for $2. It is surprising how many good mathematicians will work the problem correctly up to this last step, then forget that the check must be $2 instead of $4.

5. Ticktacktoe on the Tri-Hex pattern [*see Figure 22*] is a win for the first player, but only if he plays first on one of the black spots. Regardless of his opponent's choice of a spot, the first player can always play so that his opponent's next move is forced, then make a third play that threatens a win on two rows, thereby ensuring a win on his last move.

If the opening move is on a corner of the board, the second player can force a draw by seizing another corner. If the opening move is on a vertex of the central equilateral triangle, the second player can force a draw by taking another corner of that

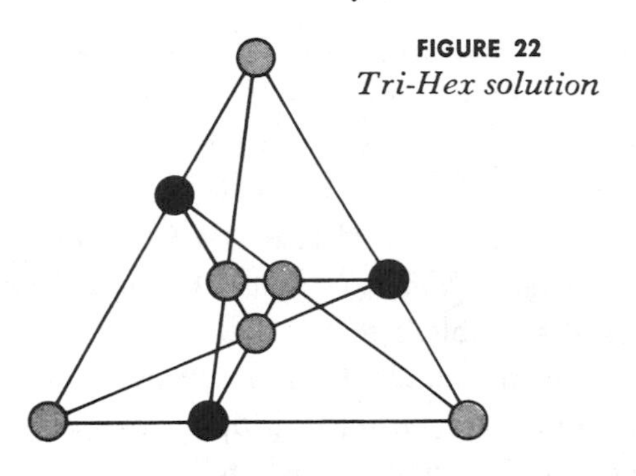

FIGURE 22
Tri-Hex solution

triangle. For a more complete analysis see "New Boards for Old Games," by Thomas H. O'Beirne, in *New Scientist*, January 11, 1962.

6. The unique solution to Langford's problem with four pairs of cards is 41312432. It can be reversed, of course, but this is not considered a different solution.

If n is the number of pairs, the problem has a solution only if n is a multiple of 4 or one less than such a multiple. C. Dudley Langford posed his problem in *The Mathematical Gazette* (Vol. 42, October 1958, page 228). For subsequent discussion see C. J. Priday, "On Langford's Problem (I)," and Roy O. Davies, "On Langford's Problem (II)," both in *The Mathematical Gazette* (Vol. 43, December 1959, pages 250–55).

The 26 solutions for $n = 7$ are given in *The Mathematical Gazette* (Vol. 55, February 1971, page 73). Numerous computer programs have confirmed this list, and found 150 solutions for $n = 8$. E. J. Groth and John Miller independently ran programs which agreed on 17,792 sequences for $n = 11$, and 108,144 for $n = 12$.

R. S. Nickerson, in "A Variant of Langford's Problem," *American Mathematical Monthly* (Vol. 74, May 1967, pages 591–95), altered the rules so that the second card of a pair, each with value k, is the kth card after the first card; put another way, each pair of value k is separated by $k - 1$ cards. Nickerson proved that the problem was solvable if and only if the number of pairs is equal to 0 or 1 (modulo 4). John Miller ran a program which found three solutions for $n = 4$ (they are 11423243, 11342324, and 41134232); five solutions for $n = 5$; 252 solutions for $n = 8$; and 1,328 for $n = 9$.

Frank S. Gillespie and W. R. Utz, in "A Generalized Langford Problem," *Fibonacci Quarterly* (Vol. 4, April 1966, pages 184–86), extended the problem to triplets, quartets, and higher sets of cards. They were unable to find solutions for any sets higher than pairs. Eugene Levine, writing in the same journal ("On the Generalized Langford Problem," Vol. 6, November 1968, pages 135–38), showed that a necessary condition for a solution in the case of triplets is that n (the number of triplets) be equal to -1, 0, or 1 (modulo 9). Because he found solutions for $n = 9$, 10, 17, 18, and 19, he conjectured that the condition is also sufficient when n exceeds 8. The nonexistence of a solution for $n = 8$ was later confirmed by a computer search.

Levine found only one solution for $n = 9$. I know of no other solution; perhaps it is unique. Readers may enjoy finding it. Take from a deck all the cards of three suits which have digit values (ace through nine). Can you arrange these 27 cards in a row so that for each triplet of value k cards there are k cards between the first and second card, and k cards between the second and third? It is an extremely difficult combinatorial puzzle.

D. P. Roselle and T. C. Thomasson, Jr., "On Generalized Langford Sequences," *Journal of Combinatorial Theory* (Vol. 11, September 1971, pages 196–99), report on some new nonexistence theorems, and give one solution each for triplets when $n = 9$, 10, and 17. So far as I am aware, no Langford sequence has yet been found for sets of integers higher than three, nor has anyone proved that such sequences do or do not exist.

7. To solve the problem of the overlapping squares, extend two sides of the large square as shown by the dotted lines in Figure 23. This obviously divides the small square into four congruent parts. Since the small square has an area of nine inches, the overlap (shaded) must have an area of 9/4, or 2¼ inches. The amusing thing about the problem is that the area of over-

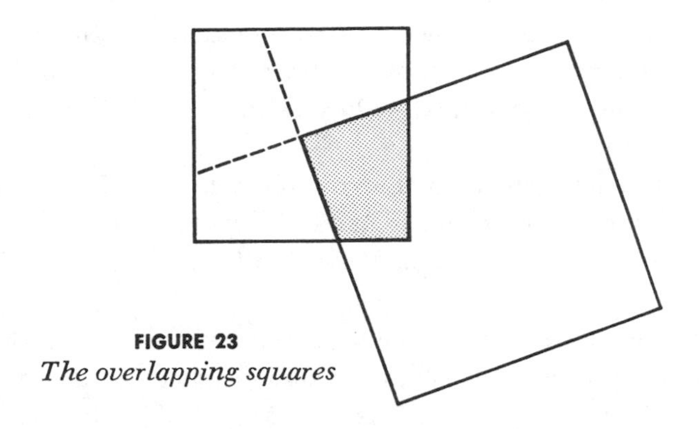

FIGURE 23
The overlapping squares

lap is constant regardless of the large square's position as it rotates around D. The fact that B trisects AC is irrelevant information, designed to mislead.

The problem appears as No. 52 in *Mathematical Challenges: Selected Problems from the Mathematics Student Journal*, edited by Mannis Charosh (National Council of Teachers of Mathematics, 1965). A second solution is given, and the problem is generalized for any regular polygon.

8. The first round of births in Fertilia produces n children, where n is the total number of mothers over as long a period as desired. The second round produces $n/2$ children, the third round produces $n/4$, and so on. The total number of children is $n(1 + \frac{1}{2} + \frac{1}{4} + \frac{1}{8} + \ldots) = 2n$. Dividing this by n gives the average number of children in a family, namely two.

Many readers pointed out that the question could be answered more simply. After showing that the ratio of boys to girls always remains one to one, it follows that in the long run there will be as many boys as there are girls. Since each family has exactly one son, there will be, on the average, also one girl, making an average family size of two children.

9. If "Oct." is taken as an abbreviation for "octal" and "Dec." as an abbreviation for "decimal," then 31 (in base-8 no-

tation) is equal to 25 (in base-10 notation). This remarkable coincidence is the basic clue in "A Curious Case of Income Tax Fraud," one of Isaac Asimov's tales about a club called The Black Widowers. (See *Ellery Queen's Mystery Magazine*, November 1976.)

John Friedlein observed that not only does Christmas equal Halloween, each also equals Thanksgiving whenever it falls, as it sometimes does, on Novem. 27. (27 in 9-base notation is 25 in decimal.)

Suzanne L. Hanauer established the equivalence of Christmas and Halloween by modulo arithmetic. Oct. 31 can be written 10/31 or 1,031. Dec. 25 is 12/25 or 1,225. And 1,031 = 1,225 (modulo 194).

David K. Scott and Jay Beattie independently established equality in an even more surprising way. Let the five letters in Oct. and Dec. stand for digits as follows: $O = 6$, $C = 7$, $T = 5$, $D = 8$, $E = 3$. We then decode Oct. 31 = Dec. 25 as:

$$675 \times 31 = 837 \times 25 = 20{,}925$$

Assuming that two different letters cannot have the same digit, that O and D (standing for initial digits of numbers) may be assigned any digit except zero, and that the other three letters may be assigned any digit including zero, there are 24,192 different ways to assign digits. Beattie actually programmed a computer to test all 24,192 possibilities. The program proved (what Scott had conjectured) that the above equation is unique!

10. To tie an overhand knot in the rope stretched from wrist to wrist, first push the center of the rope under the rope that circles the left wrist, as shown in Figure 24. Pass the loop over the left hand, then pull it back from under the rope circling the wrist. The loop will then be on the left arm, as shown at the right in the illustration. When the loop is taken off the arm,

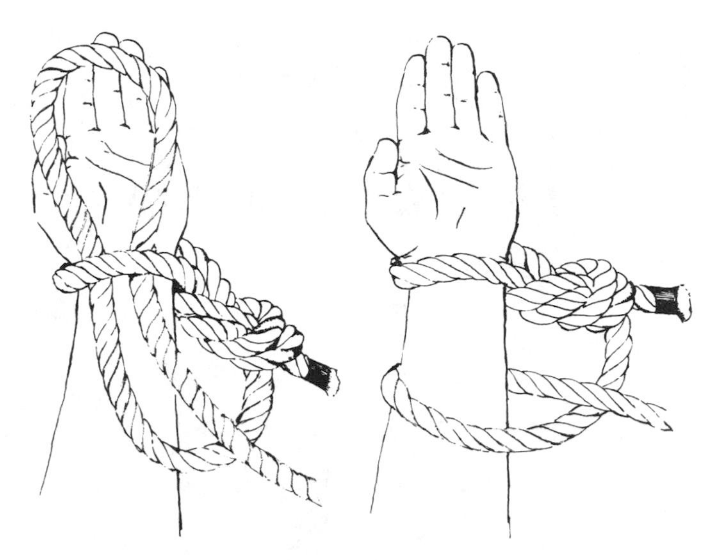

FIGURE 24
How to work the knot problem

carrying it over the left hand, it will form an overhand knot in the rope.

If the loop, after it is first pushed under the rope around the wrist, is given a half-twist to the right before it is passed over the hand, the resulting knot will be a figure eight. And if the end of the loop is pushed through a ring before the loop goes over the hand, the ring will be firmly knotted to the rope after either type of knot is formed.

Van Cunningham and B. L. Schwartz each wrote to point out that the statement of the problem did not prohibit a second solution. Fold your arms, slip one hand through one loop, the other hand through the other loop, then unfold your arms.

CHAPTER 6

Double Acrostics

IN MODERN SET THEORY two sets are said to intersect if they have one or more elements in common. The crossword puzzle, the Double-Crostic, and games of the Scrabble type can be thought of as combinatorial play in which 26 elements (letters) are arranged into sets (words) that intersect in the manner dictated by the rules and the geometric pattern on which the elements are placed. Of the hundreds of varieties of word puzzles that exploit intersection, few have been more elegant, or have had as interesting a history, as the double acrostic. It is almost forgotten today, yet it was the ancestor of the crossword puzzle and the most popular form of word play in English-speaking countries throughout the last quarter of the nineteenth century and until the end of World War I. In London in 1915 eight newspapers ran a daily double acrostic. *The World* had carried the feature since its first issue in 1874, *The Queen* since the 1860's.

The double acrostic was a highly stylized form, usually written in verse. Here, for example, is a short specimen by Tom Hood, son of the English poet Thomas Hood:

We are words that rhyme,
And we're both in time.
One is a season, the other a song;
If you guess them rightly,
you can't be wrong.

1.

It is very good fun,
If it's properly done.

2.

A beast with two toes,
How slowly it goes!

3.

The sun's overhead—
There's no more to be said!

4.

The sun's going to sink;
This is coming I think.

The first stanza gives clues for two words, called the uprights, that are spelled vertically by the initial and final letters of the words to which clues are given by the numbered stanzas. The two uprights must be of equal length and in some way related to each other. In this instance the first upright, known as the primal, is JUNE. The second upright, called the final, is TUNE. The horizontal words, defined in the numbered stanzas, are called the cross-lights or simply the lights. The complete answer is:

J e s T
U n a U
N o o N
E v E

The cross-lights could vary in length, as in this instance. It was permissible on occasion to do such things as spell them backward, use only parts of them or even scramble their letters, provided that the required operations were specified by hints in the stanzas. If the middle letters of each cross-light formed a third upright, known as the central, the puzzle was called a triple acrostic.

Henry Ernest Dudeney, in a little book called *The World's Best Word Puzzles* (1925), attributes the invention of the Victorian double acrostic to Queen Victoria! The evidence for this is in a rare volume called *Victorian Enigmas; or, Windsor Fireside Researches: Being a Series of Acrostics Enigmatically Propounded* (1861), by Charlotte Eliza Capel. Miss Capel wrote that her original double acrostics had been inspired by a puzzle given to her five years earlier by a friend at Windsor Castle who told her it had been written by the Queen for the royal children. Although the Queen's puzzle, which Miss Capel reproduced, is not in rhyme, it is certainly a double acrostic. Nine geographical words are defined (Naples, Elbe, Washington, Cincinnati, Amsterdam, Stamboul, Torneå, Lepanto, ecliptic), the initials of which spell NEWCASTLE and the finals of which, read in reverse, spell COAL MINES.

Going back five years from 1861 places the date of the Queen's puzzle at 1856. It was in the *Illustrated London News* for August 30 of that year, writes Dudeney, that the first double acrostic was printed. It bore the by-line of Cuthbert Bede, the pen name of Rev. Edward Bradley. Bradley called his puzzle an "acrostic charade" and said it was a novel type of word play "lately introduced." The clues to its uprights, LONDON and THAMES, were given by these lines:

A mighty centre of woe and wealth;
A world in little, a kingdom small.
A tainted scenter, a foe to health;
A quiet way for a wooden wall . . .

(Apparently the problem of river pollution is not a new one.)

The term "double acrostic" was firmly established by 1860, when puzzles of that type were included in a book, *Charades,*

Enigmas and Riddles, collected by someone calling himself A. Cantab. "The Double Acrostic," Cantab writes, "is a very recent invention." By 1884 the double acrostic craze had reached such proportions in England that an *Acrostic Dictionary* of 256 pages was compiled by Phillippa Pearson, whose husband wrote a number of puzzle books. It listed 30,000 common words, alphabetized by first and last letters.

Of the hundreds of thousands of double acrostic poems published in Victorian newspapers, magazines, and books, none were more charmingly written than those by Thomas Hood, the younger. Like his father, he was a skillful and prodigious writer of humorous verse, much of it published in *Fun*, a comic weekly he edited, and later in his own periodical, *Tom Hood's Comic Annual*. He turned out hundreds of puzzle poems of all types. The double acrostic in Figure 25 is from a children's book, *Excursions into Puzzledom*, by Hood and his sister, published in 1879, five years after Hood's death. Can the reader discover its uprights and cross-lights?

Ruler of all things, for a space his hand
Is traced in sparkling lines throughout the land:
Painting each pane and jewelling each tree,
Checking the brooks and rills that trickled free;
Tasting the roots and fruits all stored away,
Withering the garden blooms that were so gay.
Such is my First,—the boys alone delight
To see his silent traces over night.
And greet him well, for long they all have reckoned
Upon his aid to help them to my Second.

1

Where the fairies come, we grow,
Their most secret haunts we know.
Our fringed fans are tall and green,
Pavilions for the elfin queen.
Those that with all careful heed,
Sow at night our mystic seed,
May her sportive revels see
Underneath the greenwood tree!

[CONTINUED]

2

When a frisky fancy takes
The jovial Land of Cakes,
She calls for her piper to play her a tune,
Till the very roof-tree shakes!
And then ere it grows too late,
A perplexing figure of eight
Is danced by the lads and lasses all
At a most astonishing rate!

3

When Pierre meets Marie in the lane,
And slyly steals a kiss,
He asks a question clear and plain,
To which she answers—this!

4

No traveller of modern times
Such wondrous tales narrated;—
As of this ancient mariner
Have been most gravely stated.

5

When the storm king rises
From his cloudy lair,
And his muttered anger
Grumbles in the air;
Doors and windows rattle,
Sign-posts creak and groan.
And from roof and rafter
This is roughly blown.

FIGURE 25
A typical double acrostic by Tom Hood

It is no surprise to learn that Lewis Carroll, who enjoyed all kinds of puzzles, was addicted to the double acrostic and was the creator of many splendid specimens. His best is one that first appeared in his 1869 book *Phantasmagoria* [*see Figure 26*]. An entry in Carroll's *Diary* for June 25, 1867, reads: "Blore [one of Carroll's mathematics students at Christ Church, Oxford] brought his niece Miss Keyser to see photographs—I took a couple of her as well. Sat up listening to the music of the Christ Church Ball and wrote, at Miss Keyser's request, another one of those acrostic ballads of which I had given Blore some before."

This is how Carroll later introduced the ballad in *Phantasmagoria:* "[It] was written at the request of some young friends, who had gone to a ball at an Oxford Commemoration—and also as a specimen of what might be done by making the Double Acrostic *a connected poem* instead of what it has hitherto been, a string of disjointed stanzas, on every conceivable subject, and about as interesting to read straight through as a page of a Cyclopedia. The first two stanzas describe the two main words, and each subsequent stanza one of the cross-lights."

Carroll did not disclose the solution, but in 1932, when Macmillan issued *The Collected Verse of Lewis Carroll* (there is a Dover reprint called *The Humorous Verse of Lewis Carroll*), an anonymous editor appended to the poem what he believed to be its two uprights, COMMEMORATION and MONSTROSITIES, without supplying the cross-lights. The answer, which has dogged the poem ever since, is unquestionably wrong. As far as I know, the first printing of the correct uprights was in "The Best Acrostics," an article by H. Cuthbert Scott in *The Strand Magazine* (Vol. 50, December 1915, pages 722–28), with answers on page 109 of the next issue. The primal upright is QUASI-INSÂNITY, the final COMMEMORATION. Readers should enjoy searching for the cross-lights; there is little doubt about any but the fourth and the ninth.

It is easy to see how the double and triple acrostic, with its two or three vertical words, evolved into more complicated forms, including the crossword puzzle and such later variants as the Double-Crostic. The first crossword deserving the name was constructed by Arthur Wynne of Liverpool, who came to the United States around the turn of the century to begin a career in journalism. He was the editor of *Fun*, a Sunday supplement of the *New York World*, when he published in it on December 21, 1913, his first "Word-Cross Puzzle." (The interested

THERE was an ancient City, stricken down
With a strange frenzy, and for many a day
They paced from morn to eve the crowded town,
And danced the night away.

I asked the cause: the aged man grew sad:
They pointed to a building gray and tall,
And hoarsely answered "Step inside, my lad,
And then you'll see it all."

1

Yet what are all such gaieties to me
Whose thoughts are full of indices and surds?
$x^2 + 7x + 53$
$= \frac{11}{3}$

2

But something whispered "It will soon be done:
Bands cannot always play, nor ladies smile:
Endure with patience the distasteful fun
For just a little while!"

3

A change came o'er my Vision—it was night:
We clove a pathway through a frantic throng:
The steeds, wild-plunging, filled us with affright:
The chariots whirled along.

4

Within a marble hall a river ran—
A living tide, half muslin and half cloth:
And here one mourned a broken wreath or fan,
Yet swallowed down her wrath;

5

And here one offered to a thirsty fair
(His words half-drowned amid those thunders tuneful)
Some frozen viand (there were many there),
A tooth-ache in each spoonful.

FIGURE 26

Lewis Carroll's most difficult double acrostic

6

There comes a happy pause, for human strength
Will not endure to dance without cessation;
And every one must reach the point at length
Of absolute prostration.

7

At such a moment ladies learn to give,
To partners who would urge them overmuch,
A flat and yet decided negative—
Photographers love such.

8

There comes a welcome summons—hope revives,
And fading eyes grow bright, and pulses quicken:
Incessant pop the corks, and busy knives
Dispense the tongue and chicken.

9

Flushed with new life, the crowd flows back again:
And all is tangled talk and mazy motion—
Much like a waving field of golden grain,
Or a tempestuous ocean.

10

And thus they give the time, that Nature meant
For peaceful sleep and meditative snores,
To ceaseless din and mindless merriment
And waste of shoes and floors.

11

And One (we name him not) that flies the flowers,
That dreads the dances, and that shuns the salads,
They doom to pass in solitude the hours,
Writing acrostic-ballads.

12

How late it grows! The hour is surely past
That should have warned us with its double knock?
The twilight wanes, and morning comes at last—
"Oh, Uncle, what's o'clock?"

13

The Uncle gravely nods, and wisely winks.
It *may* mean much, but how is one to know?
He opes his mouth—yet out of it, methinks,
No words of wisdom flow.

reader will find it reproduced in Clark Kinnaird's *Encyclopedia of Puzzles and Pastimes*, 1946, page 80.) It was such an immediate success that Wynne began composing similar puzzles in all shapes and sizes.

In 1924 two young men, Richard Simon and Max Schuster, opened a book-publishing office in New York. Simon's aunt had a sick friend who was addicted to the *World*'s crosswords. Was there a book of such things, she asked her nephew, that she could give her friend? There was not. Simon and Schuster made arrangements to reprint fifty of the *World*'s crosswords ("The worst idea since prohibition," the *World*'s editors said) and persuaded the Venus Pencil Company to donate 50,000 pencils to be attached to the book's cover as a promotional stunt. The first 50,000 copies of *The Cross Word Puzzle Book*, under the imprint of the Plaza Publishing Company, sold out in three months, touching off a craze that spread quickly to England and France ("*les mots croisés*") and other countries. In Canada a bilingual form became popular, with English words going one way and French words the other. During the next twenty years the house of Simon and Schuster sold more than two million copies of crossword books.

Most newspapers in the United States began to publish a daily crossword in 1924. The astonished *World* announced late that year that its first daily puzzle would be composed by Gelett Burgess, of purple-cow fame, and added:

The fans they chew their pencils,
The fans they beat their wives,
They look up words for extinct birds—
They lead such puzzling lives!

The New York Times was the last major paper to succumb. In 1942 it instituted its Sunday puzzle, edited by Margaret Farrar, the wife of the publisher John Farrar and crossword puzzle editor of the *Times* until her retirement in 1969. (As Margaret Petherbridge she had been one of the three editors of Simon and Schuster's first puzzle book.) More than 90 percent of the nation's newspapers now have a daily crossword, and millions of crossword books are sold annually. It is the most popular type of puzzle in all parts of the world except in countries, such as China and Japan, where the language does not have individual letters and therefore is not suitable for intersecting word patterns.

The Double-Crostic, in which the words of a literary quotation and the name of the author and his work are derived from words clued by cryptic definitions, was invented by a Wellesley College graduate and Brooklyn high school English teacher, Elizabeth Seelman Kingsley. Mrs. Kingsley's first Double-Crostic appeared in the *Saturday Review of Literature* in March 1934, to be followed by a large output of Double-Crostics for that magazine, other magazines, and numerous Simon and Schuster books. When she retired in 1952, the work was continued by her assistant, Doris Nash Wortman, until Mrs. Wortman's death in 1967. The term "Double-Crostic" is a registered trademark, but the puzzle form appears regularly under other names.

One of the earliest and hardest-to-compose variants on the double acrostic is the word square, which can be thought of as a kind of ultimate acrostic because every letter in it marks an intersection of two words. A set of n horizontal words, each of n letters, intersects with another set of n vertical words of n letters that are read down. The uprights are sometimes identical with the horizontals, sometimes a different set entirely. Dudeney says in his book on word puzzles that he was the first to put the definitions for such squares into verse, and he gives a number of examples. Edmund Wilson, who shared with his literary antagonist Vladimir Nabokov a liking for word play, once tried his hand at versifying word squares. In one example from Wilson's *Night Thoughts* (1961) the five horizontal words are given first, followed by clues to five different words that form the uprights [*see Figure 27*]. Can the reader construct Wilson's square?

My *first* is a garment that fastens behind;
My *second* applies to a lush little lake;
My *third* in your *Handwörterbuch* you will find
May mean whilst or because; my *fourth* is a fake:
The Association of Impotent Old Apoplectic Parties;
My *fifth* is the steamship *Nigerian Royal Highness;*
My *sixth* a confection of musical art is;
My *seventh* an organ remote from the sinus;
My *eighth* is a painter fantastic and French;
My *ninth* is exclaimed at a wrench or a stench;
And my *tenth* is a nimble but mythical wench.

FIGURE 27
Five-by-five word-square poem by Edmund Wilson

ANSWERS

THE UPRIGHTS of Tom Hood's double acrostic are FROST and SLIDE. The cross-lights are FERNS, REEL, OUI, SINDBAD, TILE.

Lewis Carroll's double acrostic, with the uprights QUASI-INSANITY and COMMEMORATION, was given the following solution in the 1915 *Strand Magazine* article cited earlier: 1. QUADRATIC; 2. UNDERGO; 3. ALARM; 4. STREAM; 5. ICE; 6. INTERIM; 7. NO; 8. SUPPER; 9. ARENA; 10. NIGHT; 11. I; 12. TWO; 13. YAWN. The author of the article was uncertain about the ninth, suggesting ARISTA as an alternative, although admitting that neither word was satisfactory. My own guess is AURORA, which sometimes resembles folds of drapery that wave with a "mazy motion." Dmitri A. Borgmann, author of the book of word play *Beyond Language*, focuses on the phrase "tangled talk" in the ninth stanza, for which he suggests ABRACADABRA. As an alternative to the fourth cross-light, Borgmann proposes SCRIM, a coarse cotton fabric that could be "half muslin and half cloth."

The ambiguity in the fourth and ninth cross-lights brought a large number of interesting letters. Martin Burkenroad noticed that the poem contained two phrases from Coleridge's *Kubla Khan*, "river ran" and "mazy motion" as well as the capitalized word "Vision" and the cross-light "ice," but these allusions were of no help in clearing up the two ambiguous cross-lights.

I cannot list all who wrote to give their arguments, but I shall cite some of the words that were suggested. Most readers agreed that "stream" was the best choice for the fourth cross-light, referring to the divided river of men in black cloth and girls in muslin dresses. Four readers preferred "swarm," four "stoicism" ("swallowed down her wrath"), and three "schism." Other proposals were "seam," "spasm," and "scrum" (a rugby formation).

The ninth cross-light, for which "arena," "aurora," and "abracadabra" had been suggested, prompted four readers to propose "America" and four "asea." Three preferred "aphasia," three "agora." Others suggested "alfalfa," "ataxia," "arista," "arcadia," "avena," and "anarrhoea" (Greek for the flowing back of a tide after the ebb).

E. Robinson Rowe concluded, after a long and good analysis, that correct solutions for cross-lights 4 and 9 and possibly for others may have hinged on allusions to contemporary history or fiction or to local Oxford customs; the student for whom Carroll wrote his puzzle poem (he is the "Uncle" of stanzas 12 and 13) may have had much less difficulty with the answers than we do today.

The horizontal lines of Edmund Wilson's word square are APRON, REEDY, INDEM, A.I.O.A.P., and S.S.N.R.H. The vertical lines are ARIAS, PENIS, REDON, O DEAR!, and NYMPH.

CHAPTER 7

Playing Cards

PLAYING CARDS, with their numerical values, four suits, two colors, backs and fronts, and easy randomizing, have long provided recreational mathematicians with a paradise of possibilities. In this chapter we consider a few remarkable new combinatorial problems and paradoxes for which playing cards are ideal working models.

In Chapter 9 of my *New Mathematical Diversions from Scientific American,* I mentioned briefly a curious principle discovered by a young amateur magician, Norman Gilbreath. Arrange a deck of cards so its red and black cards alternate. Cut the deck to form two piles, breaking the deck so that the top cards of each pile are of different colors. If the two piles are now riffle-shuffled into each other, every pair of cards, from the top down, will consist of one red and one black card. (You can let someone else do the single shuffle and then play him 26 rounds of matching colors. You each take a card from the top of the deck; your opponent wins if the cards match. Of course you win every time.) Gilbreath later discovered that his principle is only a special case of what magicians now call the Gilbreath general principle. It applies to any repeating series of symbols and can best be explained by a few examples.

Arrange a deck so that the suits repeat throughout in the same order, say spades, hearts, clubs, and diamonds. From the top of this deck deal the cards one at a time to the table to form a pile of 20 to 30 cards. (Actually it does not matter in the least how many cards are in this pile.) Riffle-shuffle the two parts of the deck together. Believe it or not, every quartet of cards, from the top down, will now contain a card of each suit. Dozens of subtle card tricks exploiting Gilbreath's general principle have been published in magic periodicals. The simplest trick is to let someone deal and shuffle, take the deck behind your back (or under a table), then pretend to feel the suits with your fingers and bring out the cards in groups of four, each containing all four suits.

It is necessary that one packet be reversed before the shuffle. Dealing cards to the table does this automatically. Another method is to cut off a portion of the deck, turn it over and shuffle this face-up packet into the rest of the deck, which remains face down. A third method is to take cards singly from the top of the deck and push them into the pack, inserting the first card near the bottom, the next anywhere above the previously inserted card (directly above it if you wish), the third above that, and so on until you have gone as high as you can. This is equivalent to cutting off a packet, reversing its order and riffle-shuffling. The deck's original order is destroyed, of course, but the cards remain strongly ordered in the sense that each group of four cards contains all four suits.

A trick applying the Gilbreath principle to a repeating series of length 52 is to arrange one full deck so that its cards are in the same order from top to bottom as the cards in a second deck are from bottom to top. If the two decks are riffle-shuffled into each other and then cut exactly at midpoint, each half will be a complete deck of 52 different cards!

Gilbreath's general principle points up how poorly a riffle-shuffle randomizes. This inefficiency of the riffle-shuffle provided another mathemagician, Rev. Joseph K. Siberz of Boston College, with what may well be the first computer program that teaches a computer how to do a mystifying card trick. The trick uses 52 IBM punched cards, each bearing the name of a different playing card. Both program and "deck" are entered in the computer, which then prints the following instructions:

1. Give the deck several single cuts and a riffle-shuffle.
2. Cut the deck into two piles.
3. Look at the top card of one pile and remember it.
4. Bury this card in the pile from which it came, then riffle-shuffle the two piles together.
5. Cut the deck, complete the cut, and repeat several times if you wish.
6. Now give the deck back to me and I shall find your card.

If the card was, say, the five of hearts, the computer quickly prints out: "Your card was the five of hearts. Don't ask me how I do it. Magicians never reveal their secrets. Take another card and I shall do it again." If the person has failed to follow instructions exactly, the computer sometimes finds the card anyway, perhaps after asking the spectator for additional information: "I am having trouble determining the color of your card. Please help me by turning on switch *B* if it is black or switch *C* if it is red." This is followed by, "Thank you. Your card is . . ." If the instructions were not followed and the computer cannot find the card, it prints, "You did not follow my directions. Please take another card and try again." If this happens again, the computer politely asks for still another try, but after a third goof it says, "I won't find your $$=)$** card if you refuse to do it my way. Please try again."

It is not hard to see how the program finds the card. The two riffle-shuffles merely break the deck's original cyclic order into four interlocking sequences. If the instructions are followed correctly, a single card will be missing from where it should be in one of those sequences. While it is identifying the card, the computer memorizes the deck's new order and therefore is all set for an immediate repetition of the trick. The reader can easily perform the trick himself by recording the order of a deck or using an unopened pack, which comes from the manufacturer in a simply ordered sequence that can be memorized while you remove the joker and the extra cards. After a spectator has followed the instructions given above you can take the twice-shuffled deck to another room; by checking the cards off on your list it is easy to determine the single card that is out of place.

On the television show *Maverick*, popular in the mid-1960's, the gambler Bart Maverick bet someone he could take 25 cards, selected at random, and arrange them into five poker hands each of which would be a straight or better. (The hands higher than a straight are flush, full house, four of a kind, straight flush, and royal flush.) The same bet was made on television in 1967 by Paul Bryan in an episode of *Run for Your Life*. It is what gamblers call a "proposition"—a bet for which the odds seem against the person making it when actually they are strongly in his favor. If the reader will experiment with 25 randomly chosen cards, he will be surprised at the ease with which five hands can be arranged. Try the flushes first (there will be at least two), then look for straights and full houses. I have no idea of the actual probability of success, but it is extremely high. Indeed, the question arises: Is success always possible? The answer is no. There are sets of 25 cards that cannot be partitioned into five poker hands of straight or better.

With this introduction the reader is invited to consider the 25 cards shown in Figure 28. Can the bet be won with this set? If so, find five hands. If not, prove that it is impossible. This ingenious puzzle is quickly solved if you go about it correctly; a single card is the key. The puzzle was sent to me by Hamp Stevens.

For a second combinatorial problem the reader is asked to place any three playing cards face down in a row. The task is to turn one card at a time and in seven moves produce all the $2^3 = 8$ different permutations of face-up and face-down cards, ending with the three cards face up. There are six ways to do it. Letting *F* stand for the face, *B* for the back, one solution is *BBB*, *BBF*, *BFF*, *BFB*, *FFB*, *FBB*, *FBF*, and *FFF*. (These eight permutations, by the way, correspond to the eight rows of a "truth table" giving the eight possible combinations of true and false for three statements in the propositional calculus of symbolic

FIGURE 28

Can these cards be arranged to form five poker hands, all straight or better?

logic.) Is there a solution with four cards? There are now $2^4 = 16$ different permutations. The problem is to start with four face-down cards, turn one at a time, and in 15 moves run through all 16 permutations, the last one being the four cards face up. It turns out that this is not possible. The problem, which appeared in Mannis Charosh's column in *The Mathematics Student Journal*, is to find a simple way of proving impossibility. The proof leads quickly to a generalization for rows of n cards.

C. L. Baker has called my attention to a little-known solitaire game he found could be simplified to provide an endless series of engrossing combinatorial puzzles. The game was taught Baker by his father, who in turn learned it from an Englishman during the 1920's. It differs from most solitaire games in that, although the initial pattern is determined by chance, once the cards have been placed the player has complete information. Each initial layout is therefore either solvable or unsolvable, and finding a solution becomes a stimulating challenge. As in chess, one must think ahead many moves because any mistake in play is irreversible. The game has a peculiar flavor of its own, somewhat akin to sliding-block puzzles. The probability of winning is high if one is skillful, and the more one plays the more skillful one becomes.

To play the game with a full deck shuffle the 52 cards, then deal them face up into the eight-column array shown in Figure 29. The dealing is left to right, with the cards overlapping as pictured. (This places seven cards in the first four columns and six in the last four.) Baker calls the columns "board columns" and labels them *B1* to *B8*. The four dotted cells above the layout are called "playoff cells," *P1*, *P2*, *P3*, *P4*. They are empty at the start. The object of the game is to place on each of those cells, in consecutive order starting with the ace, all 13 cards of the same suit. The game is won if, and only if, all four playoff cells are completely filled.

The four dotted cells below the layout, *T1, T2, T3, T4,* are called "temporary cells." In the course of play a single card (no

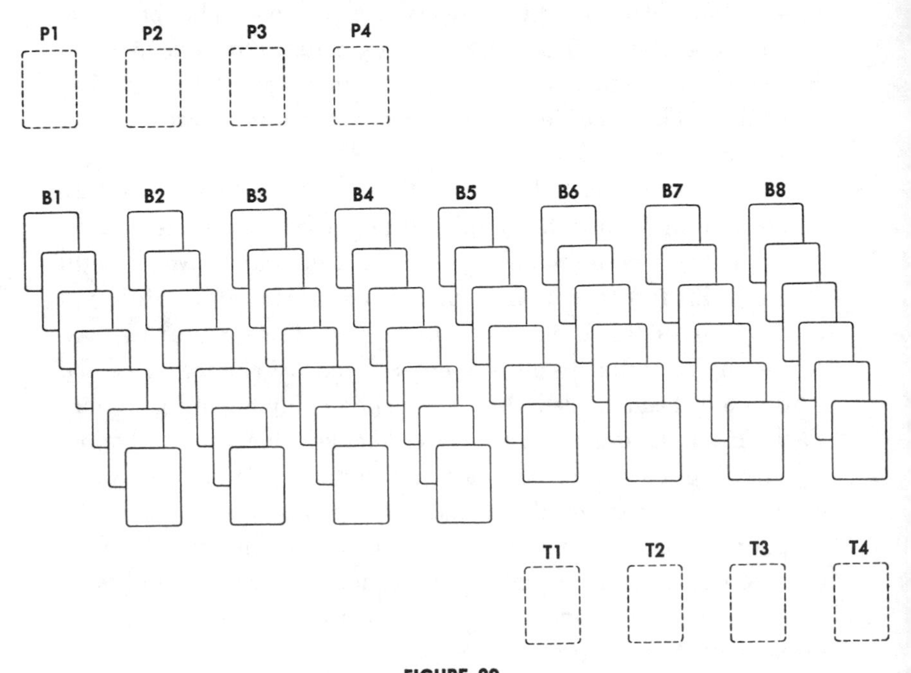

FIGURE 29
Schematic layout for C. L. Baker's solitaire game, order-4 version

more) can be placed on any one of those cells. They can, therefore, hold a maximum of four cards.

The rules are as follows:

1. Only one card may be moved at a time.

2. Only the top (uncovered) card may be moved from any *B* column.

3. An ace may be played on any empty *P* cell. On top of it may be played the deuce of the same suit, then the three, and so on through the king. These cards may come from the exposed ends of columns or from the *T* cells.

4. An uncovered card in a *B* column may be shifted to become the uncovered card of another *B* column only if it goes on a next-higher card of the *same suit.*

5. Any card may be played on an empty *T* cell and remain there until you wish to move it somewhere else.

6. If a *B* column becomes empty, any movable card may be placed on it as a new starting card.

7. A card on a *T* cell may be moved to a *P* cell, to an empty *B* column, or to a filled *B* column provided it goes on a next-higher card of the same suit.

It was Baker's happy discovery that removing one, two, or three suits from the full deck creates games of "lower order." For the three-suit, or order-3, game, the cards are dealt into seven columns, and there are three *P* cells and three *T* cells. The order-2 game (two suits) has six columns, two *P* cells, and two *T* cells. The order-1 game has five columns, one *P* cell, and one *T* cell. Are all layouts solvable?

The order-2 is an ideal introduction to the game. The reader is urged at this point to obtain a deck, remove two of the suits, shuffle the remaining 26 cards, and deal a random layout. Only actual play will convey the game's fascination. It is a good plan to keep a record of each starting pattern because if you fail to win, you may want to restore the original layout and try a different strategy. Perhaps a friend or a member of the family will take a crack at the same pattern. Not all starting layouts are solvable, but often a seemingly hopeless pattern can finally be broken by devious lines of play.

For our final problem the reader is asked to try his skill at the order-2 game shown in Figure 30, in which the spades and hearts are arranged consecutively. It can be solved, but what is the shortest solution?

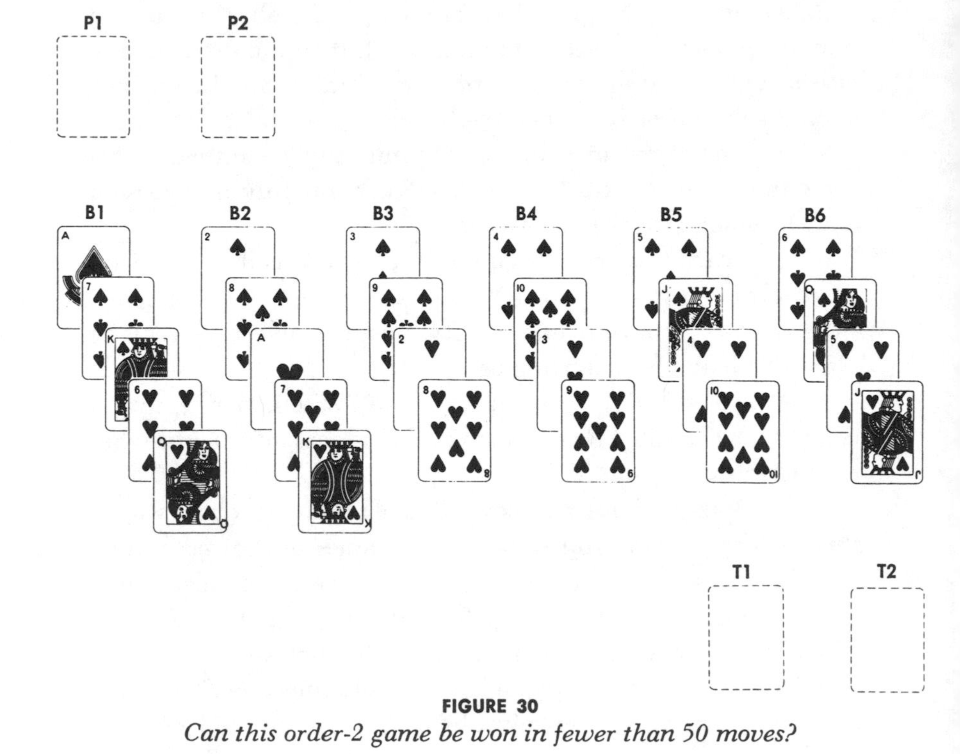

FIGURE 30

Can this order-2 game be won in fewer than 50 moves?

All kinds of difficult combinatorial questions are posed by Baker's game (I call it that because I have not so far discovered if the game has a name). Among many questions raised by Baker, and for which he has no answers, are the following: Does the probability of winning decrease as order-n increases? If so, does it approach zero or some nonzero limit? For a given order what is the maximum number of moves that may be required for a minimum-move solution?

ANSWERS

The set of 25 cards cannot be arranged to make five poker hands each of which is a straight or better. The key to the proof is the four of hearts. There is no three or five in the set, so the four of hearts cannot be part of a straight. There are only three other heart cards, so it cannot be part of a flush. It is the only four, so it cannot be part of a full house. Finally, there is no set of four cards of like value, so the four of hearts cannot be the fifth card accompanying four of a kind.

The second problem was to prove that, starting with a row of four face-down cards, you cannot turn up one card at a time and run through all 16 possible permutations of face-up and face-down cards, ending with four face-up cards. The proof uses a simple parity check. Each time a card is turned, it changes the parity of the face-up cards from odd to even or vice versa. At the start the parity of face-up cards is even (zero is an even number), therefore the 16th and last permutation must be odd. The problem specified, however, that the final permutation must be four face-up cards, an even number; consequently the problem is unsolvable.

Are all order-1 layouts of the solitaire game solvable? The answer is no. There are many thousands of impossible initial patterns (e.g., 6, J, A, 8, 9 in the first row, K, 4, 3, 7, 10 in the second, Q, 5, 2 in the third). It is estimated, however, that more than 99 percent of the starting patterns are solvable, and their solutions are easily found. A more difficult question, not yet answered, is to establish the maximum number of moves that provides a minimum solution for an initial layout. Nor has this been answered for the order-2 game.

The fourth problem was to solve the order-2 game with the starting pattern shown in Figure 30, and to do it in the fewest number of moves. More than 65 *Scientific American* readers found solutions in 49 moves, surely the minimum. All had in common the placing of 13 hearts on a *P* cell in 32 moves, with the remaining 17 moves used to put the remaining spades on

MOVE	CARD	TO	MOVE	CARD	TO
1	K ♥	T1	26	7 ♥	P1
2	8 ♥	B4	27	8 ♥	P1
3	7 ♥	B4	28	9 ♥	P1
4	A ♥	P1	29	10 ♥	P1
5	2 ♥	P1	30	J ♥	P1
6	8 ♠	B3	31	Q ♥	P1
7	2 ♠	T2	32	K ♥	P1
8	K ♥	B2	33	7 ♠	T2
9	Q ♥	B2	34	8 ♠	B1
10	6 ♥	B4	35	9 ♠	B2
11	K ♠	T1	36	3 ♠	P2
12	7 ♠	B3	37	10 ♠	B3
13	A ♠	P2	38	4 ♠	P2
14	2 ♠	P2	39	J ♠	B4
15	J ♥	B2	40	5 ♠	P2
16	10 ♥	B2	41	Q ♠	B5
17	6 ♥	T2	42	6 ♠	P2
18	7 ♥	B1	43	7 ♠	P2
19	6 ♥	B1	44	8 ♠	P2
20	8 ♥	T2	45	9 ♠	P2
21	9 ♥	B2	46	10 ♠	P2
22	3 ♥	P1	47	J ♠	P2
23	4 ♥	P1	48	Q ♠	P2
24	5 ♥	P1	49	K ♠	P2
25	6 ♥	P1			

FIGURE 31
A 49-move solution for solitaire game

the other P cell. The first 49-solution received, from Warren H. Ohlrich, is shown in Figure 31.

CHAPTER 8

Finger Arithmetic

Ah why, ye gods! should two and two make four?
—ALEXANDER POPE, The Dunciad, Book 2

ANTHROPOLOGISTS have yet to find a primitive society whose members are unable to count. For some time they assumed that if an aboriginal tribe had no words for numbers except "one," "two," and "many," its members could not count beyond two, and they were mystified by the uncanny ability of such people to look over a herd of sheep, for example, and say one was missing. Some anthropologists believed these tribesmen had a phenomenal memory, retaining in their heads a gestalt of the entire herd, or perhaps knew each sheep personally and remembered its face. Later investigators discovered that the use of the same word for all numbers above two no more meant that a tribesman was unaware of the difference between five and six pebbles than the use of the same word for blue and green meant that he was unaware of the difference in color between green grass and blue sky. Tribes with limited number vocabularies had elaborate ways of counting on their fingers, toes, and other parts of their anatomy in a specified order and entirely in their heads. Instead of remembering a word for 15 a man simply recalled he had stopped his mental count on, say, his left big toe.

Most primitive counting systems were based on 5, 10, or 20, and one of the few things on which cultural anthropologists are in total agreement (and in agreement with Aristotle) is that the reason for this is that the human animal has 5 fingers on one hand, 10 on both, and 20 fingers and toes. There have been many exceptions. Certain aboriginal cultures in Africa, Australia, and South America used a binary system. A few developed a ternary system; one tribe in Brazil is said to have counted on the three *joints* of each finger. The quaternary, or 4-base, system is even rarer, confined mostly to some South American tribes and the Yuki Indians of California, who counted on the *spaces* between their fingers.

The 5-base has been used much more widely than any other. In many languages the words for "five" and "hand" either are the same or are closely related to earlier words; *pentcha,* for example, is "hand" in Persian and *pantcha* is "five" in Sanskrit. The Tamanacos, a South American Indian tribe, used the same word for 5 that they used for "a whole hand." Their word for 6 meant "one on the other hand," 7 was "two on the other hand," and so on for 8 and 9; 10 was "both hands." For 11 through 14 they stretched out both hands and counted "one on the foot, two on the foot," and so forth, until they came to 15, which was "a whole foot." As one might guess, the system continued with 16 expressed as "one on the other foot," and so on through 19. Twenty was the Tamanacos' word for "one Indian," 21 was "one on the hand of another Indian." "Two Indians" meant 40, "three Indians" 60. The ancient Java and Aztec weeks were five days long, and there is a theory that the Roman X for 10 was derived from two V's, one upside down, and that the V was a representation of a human hand.

Early number words were frequently identical with words for fingers, toes, and other parts of the body. The present English use of "digit," from the Latin for "finger," for the ten numbers 0 through 9 testifies to an early finger origin of Anglo-Saxon counting. There are amusing exceptions. The Maori word for 4 is "dog," apparently because a dog has four legs.

Among the Abipónes, a now vanished South American Indian tribe, the word for 4 meant "the toes of the rhea"—three in front and one in back.

Primitive number systems with bases of 6 through 9 are extremely rare. Apparently once people found a need to name numbers greater than five they usually jumped from one hand to the other and adopted a 10-base system. The ancient Chinese used a base of 10, as did the Egyptians, the Greeks, and the Romans. One of the curiosities of ancient mathematics was the sexagesimal (base-60) system that the Babylonians took over from the Sumerians and with which they achieved a remarkably advanced mathematics. (Our ways of measuring time and angles are relics of the Babylonian system.) Today 10 is almost universal as a number base throughout the world, even among primitive tribes. David Eugene Smith, in the first chapter of his *History of Mathematics*, first published in 1923, reports that a survey of seventy African tribes revealed that all of them used a 10-base system.

Above 5, very few number systems have been based on primes. W. W. Rouse Ball, in *A Short Account of the History of Mathematics* (fourth edition, 1908), cites only the 7-base system of the Bolas, a West African tribe, and the 11-base system of the early Maoris, although I cannot vouch for either assertion. Vigesimal, or 20-base, systems (fingers plus toes) were fairly common, the Mayan system being the outstanding instance. Because it used both zero and positional notation it was one of the most advanced of the ancient number systems, far superior, for example, to the clumsy Roman system (a statement that gives the jitters to cultural relativists since it suggests a value judgment that vaults cultural boundaries). The 20-base system survives today as words in such languages as French (*quatre-vingts* for 80), English ("Fourscore and seven years ago . . ."), and particularly Danish, in which number names are based on a curious mixture of the decimal and vigesimal systems.

The obvious connection between 5 and 10, the most popular

ancient bases, and the fingers of one and two hands has suggested to many science-fiction writers that the number systems of extraterrestrial humanoids are likewise based on the number of fingers they possess. (The creatures in Walt Disney's animated-cartoon culture presumably use a 4- or 8-base system, since they have only four fingers on each hand.) Harry L. Nelson of Livermore, California, sent the following puzzle: Suppose a space probe to Venus sends back a picture of an addition sum scratched on a wall [*see Figure 32*]. Assuming that the Venusians use a positional notation like ours and a number base corresponding to the fingers on one Venusian hand, how many fingers are on that hand? (We also assume that numbers do not begin with zero. Otherwise, the sum could be in decimal notation: 05 + 05 = 010.)

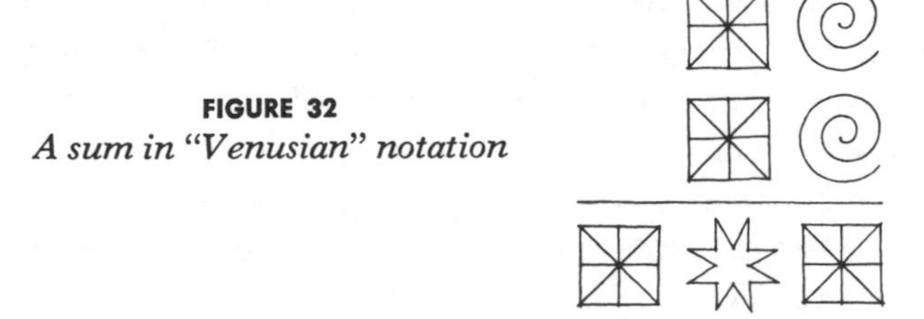

FIGURE 32
A sum in "Venusian" notation

Now that the decimal system is so universally established there seems to be no chance that the human race will convert to another number base, in spite of the fact that a duodecimal (base-12) system offers certain practical advantages, such as having four divisors for the base compared with only two for the decimal base. It has had enthusiastic advocates for centuries. And there are technical advantages, although mostly for number theorists, to a prime-number base, such as 7 or 11, as argued by the eighteenth-century French mathematician Joseph Louis Lagrange.

Powers of 2, particularly 8 and 16, have been defended as number bases by many mathematicians. "As there is no doubt that our ancestors originated the decimal system by counting on their fingers," wrote W. Woolsey Johnson in the *Bulletin of the New York Mathematical Society* (October 1891, page 6), "we must, in view of the merits of the octonary system, feel profound regret that they should have perversely counted their thumbs, although nature has differentiated them from the fingers sufficiently, she might have thought, to save the race from this error."

Donald E. Knuth has discovered that Emanuel Swedenborg, in 1718, wrote a treatise, *A new system of reckoning which turns on 8 instead of the usual turning at 10*, translated by Alfred Acton and published by the Swedenborg Scientific Association, Philadelphia, in 1941. Swedenborg gives a new nomenclature for digits and concludes, "Should the practice of the use and the use of the practice give its approval, I suppose that the learned world will gain incredible benefits from this octonary reckoning." Incidentally, it has recently been discovered that crows are capable of counting up to 7. See "The Brain of Birds," by Laurence Jay Stettner and Kenneth A. Matyniak, *Scientific American*, June 1968.

In a chapter on the ternary system (in my *Sixth Book of Mathematical Games*) I mentioned the strange nomenclature devised by two mathematicians who preferred a 16-base. I hasten to add that modern computers have long been using a base-8 arithmetic; more recently a "hexadecimal" (base-16) arithmetic, using the 16 digits 0, 1, 2, 3, 4, 5, 6, 7, 8, 9, A, B, C, D, E, F, has become an important part of the language of IBM's System/360 computers.

Just as primitive societies varied in their choice of a number base, so they varied in the style in which they counted. Since most people are right-handed, counting was usually started on the left hand, sometimes in an unvarying, ritualistic way and sometimes not. A person might begin the count at the thumb or little finger, either by tapping with a right finger, by bending down the left fingers, or by starting with a closed fist and opening the fingers one at a time. On the Andaman Islands in the Bay of Bengal people started with the little finger and tapped their nose with successive fingers. On an island in the Torres Strait between Australia and New Guinea people would count to five by tapping the fingers of their left hand, but instead of going on to the right hand they tapped their left wrist, left elbow, left shoulder, left nipple and sternum, then continued the count by reversing this order on the right side of their body. Mathematicians have made the point that when fingers and other parts of the body are successively tapped in counting, they are being used to express ordinal numbers (first, second, third, and so on), whereas when fingers are raised all at once to signify, say, four frogs, they are expressing the cardinal number (one, two, three, and so on) of a set.

The ancient Greeks had an elaborate hand symbolism for counting from one to numbers in the thousands; it is mentioned by Herodotus but little is known about its finger positions. The ancient Chinese and other Oriental cultures had finger symbols of similar complexity that are still used for bargaining in bazaars, where the expressed number can be concealed from bystanders by a cloak. The Roman method of symbolizing numbers with the hands is mentioned by many Roman authors. In the eighth century the Venerable Bede devoted the first chapter of a Latin treatise, *The Reckoning of Times* (such as calculating the dates of Easter), to a Roman system of finger symbols that he extended to one million. (His symbol for one million is the clasping of both hands.)

Most arithmetic manuals of the medieval and Renaissance periods included such methods. A typical system, shown in Figure 33, is from the first important mathematical book to be printed, a 1494 Italian work by Luca Pacioli, a Franciscan friar (who later wrote a book on the golden ratio that was illustrated by his friend Leonardo da Vinci). The Roman poet Juvenal had such a system in mind when he wrote in his *Satires:* "Happy is he indeed who . . . finally numbers his years upon his right hand"; that is, happy is he who lives to be 100, the number at which the right hand was first used in the symbolism. Note that most of the left-hand symbols have right-hand duplicates and that even on the same hand certain symbols seem to be the same unless there are subtle differences not made clear in the crude drawings. St. Jerome wrote in the fourth century that 30 was associated with marriage, the circle formed by the thumb

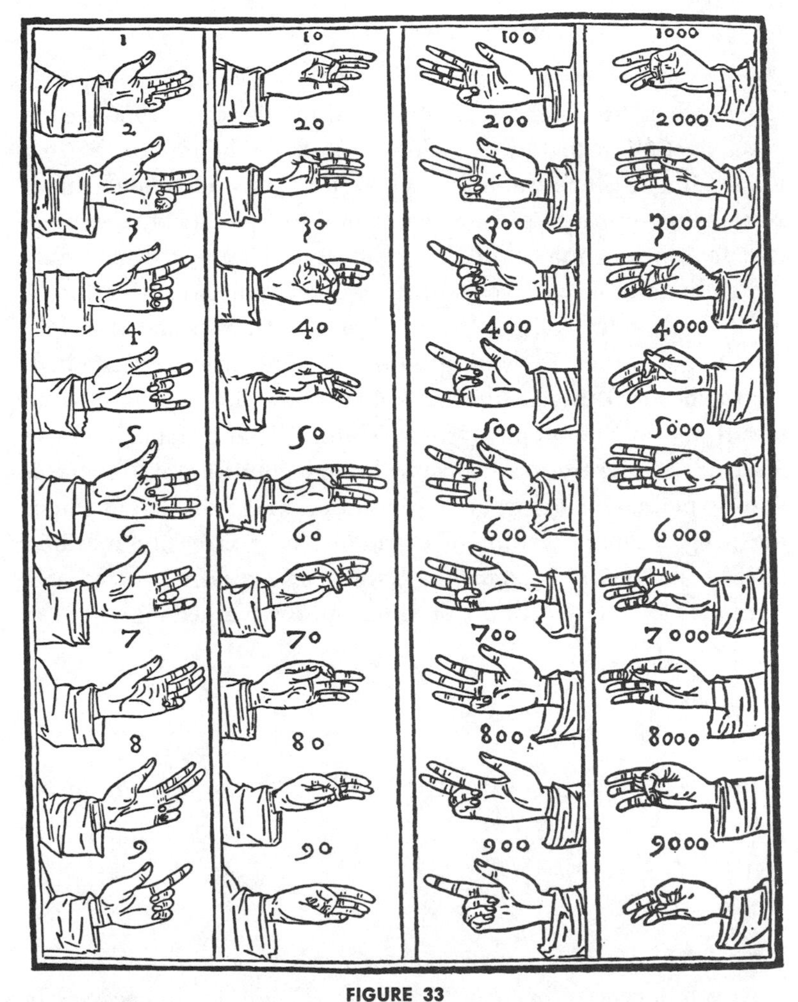

FIGURE 33

Italian finger symbolism as illustrated by Luca Pacioli in 1494

and first finger symbolizing the union of husband and wife; similarly, 60 was associated with widowhood, symbolized by the breaking of this circle.

All these old methods of finger symbolism have a 10-base, but there is no reason why fingers cannot be used just as easily for counting in systems with other bases. Indeed, the fingers are peculiarly adapted to the simplest of all systems, the binary, since a finger raised or lowered is comparable to a flip-flop circuit in modern computers that use binary counting. Frederik

Pohl, in a magazine article, "How to Count on Your Fingers," reprinted in his *Digits and Dastards* (Ballantine, 1966), suggests starting with the fists closed, backs of hands up. An extended finger is one in the binary system, an unextended finger zero. Thus in order to count from one to 1111111111 (equivalent to 1,023 in the decimal system) one begins by extending the right little finger. To indicate the decimal two, which is 10 in the binary system, the little finger is retracted and the right ring finger raised. Extending both ring and little fingers yields 11, the decimal three. Figure 34 shows how the two hands represent 500 in the binary system. With a little practice one can learn to use the fingers for rapid binary counting and even, as Pohl explains, for doing binary addition and subtraction. Since the propositional calculus of symbolic logic is easily manipulated in the binary system, the hands can actually be used as a computer for solving simple problems in two-valued logic.

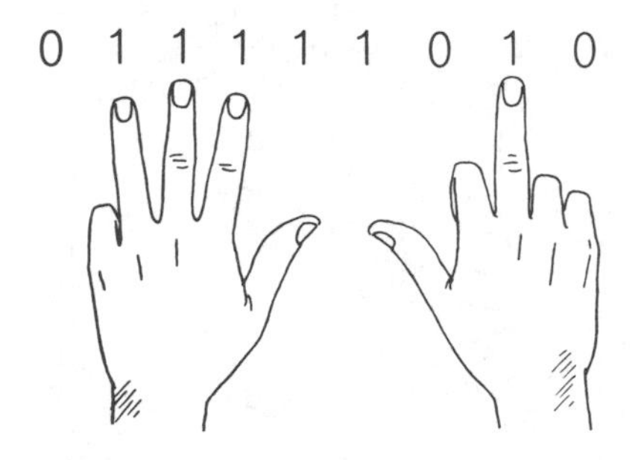

FIGURE 34
Binary 500 shown by the fingers

Any binary number consisting entirely of 1's is necessarily one less than a power of 2; the number 1,023, for example, indicated in binary digits by extending all 10 fingers, is $2^{10} - 1$. This suggested an interesting puzzle to Pohl. Suppose we wish to subtract a certain number n from 1,023 (or any lower number expressed in binary as a string of 1's). Can you think of an extremely simple way to perform such a subtraction quickly with the fingers?

Since few people in the Middle Ages and the Renaissance learned the multiplication table beyond 5×5 or had access to an abacus, a variety of simple methods were in use for obtaining the products of numbers from 6 through 10. One common method, called "an ancient rule" in a 1492 book, was to use the complements of the two numbers with respect to 10. (The complement of n would be $10 - n$.) To multiply 7 by 8, write down their complements, 3 and 2. Either complement, taken from the number with which it is *not* paired, gives 5, the number of 10's in the product of 7 and 8. The product of 3 and 2 is 6. Fifty added to 6 is 56, the final answer.

The fingers of both hands were often used as a computing device for this method. On each hand the fingers are assigned numbers from 6 through 10, starting with the little fingers. To multiply 7 and 8, touch the 7 finger of either hand to the 8 finger of the other, as shown at the top of Figure 35. Note that the complement of 7 is represented by the three upper fingers (those above the touching fingers) of the left hand, and the complement of 8 by the two upper fingers of the right hand. The five lower fingers represent 5, the number of 10's in the answer. To 50 is added the product of the upper fingers, 2×3, or 6, to obtain 56. This simple method of using fingers to compute the product of any pair of numbers in the half-decade 6 through 10 was widely practiced during the Renaissance and is said to be used still by peasants in parts of Europe and Russia.

The method has considerable pedagogical value today in the elementary schools, not only because children are intrigued by it but also because it ties in neatly with the algebraic multiplication of binomials. Instead of using complements up to 10, we can best represent 7 and 8 as excesses over 5, writing them as the binomials $(5 + 2)$ and $(5 + 3)$, then performing the multiplication:

$$\begin{array}{l} 5 + 2 \\ \underline{5 + 3} \\ 25 + 10 \\ \underline{\quad + 15 + 6} \\ 25 + 25 + 6 = 56. \end{array}$$

The first two numbers on the lowest line correspond to the sum of the lower fingers multiplied by 10; the 6 corresponds to the product of the upper fingers.

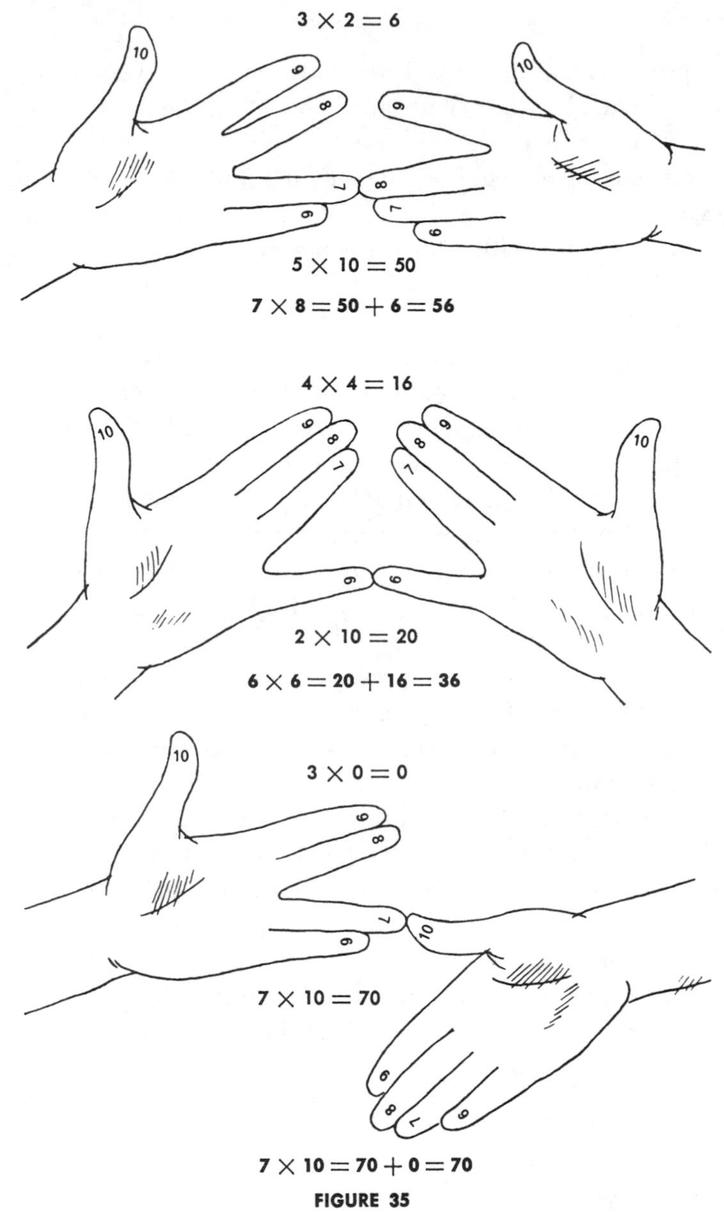

FIGURE 35

How fingers multiply pairs in the half-decade 6 through 10

The finger method of multiplying generalizes easily to half-decades higher than 10, although there is no evidence it was ever used for numbers beyond 10. For all half-decades ending in 5 a slightly different procedure is used. Consider the next-higher half-decade, 11 through 15, and suppose we wish to multiply 14 and 13. The fingers are assigned numbers from 11 through 15, and the fingers representing the numbers to be multiplied are touched as shown in Figure 36. The seven lower fingers are multiplied by 10 to obtain 70. Now, however, instead of adding the product of the upper fingers, we ignore the upper fingers and obtain the product of the two sets of lower fingers, 4×3, or 12. Adding this to 70 yields 82. The final step is to add the constant 100. This gives the final answer, 182.

There are many ways to explain why this works, but the simplest is to think in terms of binomial multiplication:

$$\begin{array}{r} 10 + 3 \\ \underline{10 + 4} \\ 100 + 30 \\ \underline{+ 40 + 12} \\ 100 + 70 + 12 = 182. \end{array}$$

The 100 on the left is the additive constant, 70 is the sum of the lower fingers multiplied by 10, and 12 is the product of the two sets of lower fingers.

For all half-decades ending in 0 we return to the first procedure. For 16 through 20 each lower finger has a value of 20 and the additive constant jumps to 200, as in the multiplication

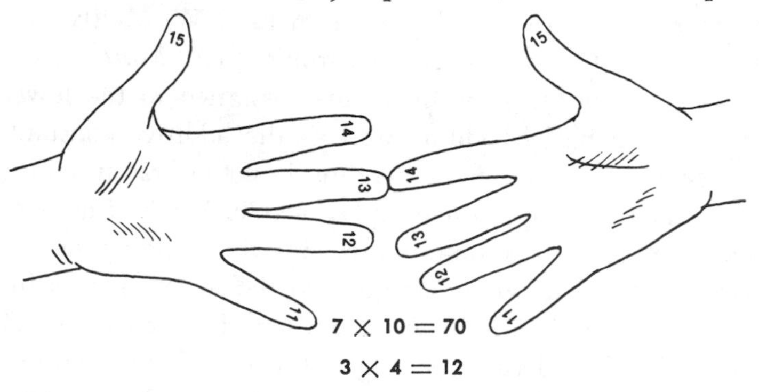

$13 \times 14 = 70 + 12 + 100 = 182$

FIGURE 36

Multiplying in the half-decade 11 through 15

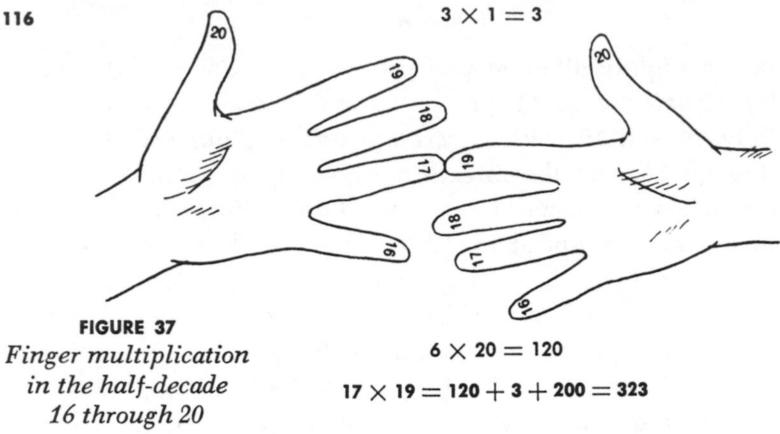

FIGURE 37
Finger multiplication in the half-decade 16 through 20

of 17 and 19 [*see Figure 37*]. Multiplying the six lower fingers by 20 gives 120. The product of the two sets of upper fingers yields 3. To 123 we add the constant, 200, to get 323, the final answer. The binomial schemata is

$$\begin{array}{l} 10 + 7 \\ \underline{10 + 9} \\ 100 + 70 \\ \underline{\quad\;\; + 90 + 63} \\ 100 + 160 + 63 = 323. \end{array}$$

If we move the 100 in 160 to the left and the 60 in 63 to the middle, we have 200 + 120 + 3. This corresponds to the finger computation. The constant is 200, the sum of the lower fingers times 20 is 120, and the product of the upper sets of fingers is 3.

The chart in Figure 38, adapted from Ferd W. McElwain's article "Digital Computer—Nonelectronic" (in *Mathematics Teacher*, April 1961), gives the values assigned to the lower fingers for each half-decade as well as the additive constant. Remember, for each half-decade ending in 0 the first system is used, in which the upper fingers play a role. For half-decades ending in 5 the second system is used, in which the upper fingers are ignored. The value assigned to the lower fingers for half-decades ending in 5 is $10(d-1)$, where d is the number of the decade. For half-decades ending in 0 it is $10d$. The additive constant for half-decades ending in 5 is $100(d-1)^2$; for half-decades ending in 0 the constant is $100d(d-1)$.

The chart extends to all higher half-decades. There are many ways to write general formulas that cover the entire procedure.

Decade	Half-Decades	Value of Lower Fingers	Additive Constant
1	1-5	0	0
	6-10	10	0
2	11-15	10	100
	16-20	20	200
3	21-25	20	400
	26-30	30	600
4	31-35	30	900
	36-40	40	1,200
5	41-45	40	1,600
	46-50	50	2,000

FIGURE 38
Chart showing finger values and constants for multiplying numbers up to 50

Nathan Altshiller Court, in *Mathematics in Fun and in Earnest* (Dial, 1958), gives the following:

$$(a + x)(a + y) = \\ 2a(x + y) + (a - x)(a - y),$$

which can also be written

$$(a + x)(a + y) = a(x + y) + xy + a^2,$$

where x and y are the final digits of the numbers to be multiplied and a can be 5, 10, 15, 20, 25, 30, . . . , the first numbers of each half-decade.

Can this finger-computing system be adapted to the multiplication of numbers from different half-decades, say 17×64? The answer is yes. The procedure is complicated, unfortunately, requiring the assignment of different values to fingers of each hand, so that I must refer the reader to McElwain's article cited above, in which a method is explained. Of course, one can always break the larger number into smaller parts for a series of finger multiplications that are then added to get the final result. Thus 9×13 can be obtained by adding (9×6) to (9×7).

There is a philosophical insight in all of this. Pure mathematics, in one obvious sense, is a construction of the human mind, but there also is an astonishing fit between pure mathematics and the structure of the world. The fit is particularly close with respect to the behavior of physical objects, such as pebbles and fingers, that maintain their identities as units. Thus $2 + 2 = 4$ is both a law of pure arithmetic, independent of the actual world, and a law of applied arithmetic. Every now and then a cultural anthropologist, overeager to drag science and mathematics into the folkways, argues that because different tribes have calculated with different number systems, mathematical laws are entirely cultural, like traffic regulations and baseball rules. He forgets that different base systems for the natural numbers are no more than different ways of symbolizing and talking about the *same* numbers, and are subject to the same arithmetical laws regardless of whether the number manipulator is a Harvard mathematician or an aborigine adding on his fingers.

The plain fact is that there is no place on the earth or on any other planet where two fingers plus two fingers is anything but four fingers. The only exception I have come on is in George Orwell's *1984*, in that terrible torture scene in which Winston Smith is finally persuaded that two plus two is five:

> *O'Brien held up the fingers of his left hand, with the thumb concealed.*
>
> *"There are five fingers there. Do you see five fingers?"*
>
> *"Yes."*
>
> *And he did see them, for a fleeting instant, before the scenery of his mind changed. He saw five fingers, and there was no deformity.*

The same possibility had been raised by Dostoevski. "Mathematical certainty is, after all, something insufferable," says the narrator of *Notes from Underground.* "Twice two makes four seems to me simply a piece of insolence. Twice two makes four is a pert coxcomb who stands with arms akimbo barring your path and spitting. I admit that twice two makes four is an excellent thing, but if we are to give everything its due, twice two makes five is sometimes a very charming thing too."

Charming, perhaps, but applying to no logically possible world. It is a subjective, self-contradictory delusion, one that can be temporarily induced only by a "collective solipsism" (as Orwell called it) in which all truth, including scientific and mathematical truth, is defined without reference to the abstract laws of logic or to the mathematical patterns of the external world.

ADDENDUM

J. A. LINDON, who lives in Addlestone, England, is in my opinion the greatest living writer in English of humorous verse. Since there is almost no market for his work (the one exception in the United States is *The Worm Runner's Digest*), most of his poems are written to send to friends who, I trust, have the sense to preserve them. The following poem, on the first discovery of arithmetic, was sent to me in 1968 shortly after this chapter had appeared in *Scientific American.*

FOUNDATIONS OF ARITHMETIC

BY J. A. LINDON

One day when Mugg the Missing Link was prowling through the woods,
In search of wives and mammoth-meat and other useful goods,
Whom should he see, on pushing out from deep arboreal shade,
But Ogg the Paleolithic Man, cross-legged in a glade.

This Ogg had made a neat array of pebbles on the ground,
In number they were twenty-one, the most that could be found,
And Ogg, with one red-hairy hand pressed to his bony brow,
Was staring at these pebbles like a ruminating cow.

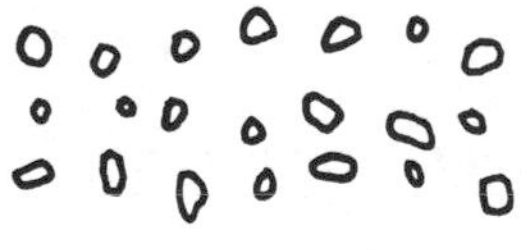

Thought Mugg—for he was Primitive—I should be very dull
To lose this opportunity of busting in his skull;
My club weighs half a hundredweight, he doesn't wear a hat—
(And here he wondered) Yes, but what the Devil is he at?

For Ogg was touching pebbles and then prodding at his digits,
Until the weirdness of it all afflicted Mugg with fidgets:
"Invented any goodish wheels just recently?" he hollered,
And doubled up in merriment, his face raw-beefy coloured.

Ogg looked at him in pity, then he drummed upon his chest,
His reddish eyes aflame with all a mathematician's zest:
"I've done a Think!" he bellowed. "Monkey Mugg, I've done a Think!
And I would write it down, but no one's yet invented ink."

Mugg moved a little closer, and his eyes and mouth were round,
And stared in trepidation at those pebbles on the ground.
Ogg pointed with a nailed red-hairy sausage at the rows
And said, "Three people's hand-plus-two is hands-plus-feet-plus-nose."

"And this is hand-plus-two of people's three-for-each-by-name,
So three times hand-plus-two and hand-plus-two times three's the same!"
Mugg scratched his matted hairy head, not knowing what to say.
Said Ogg, "It's all made clear by this rectangular array."

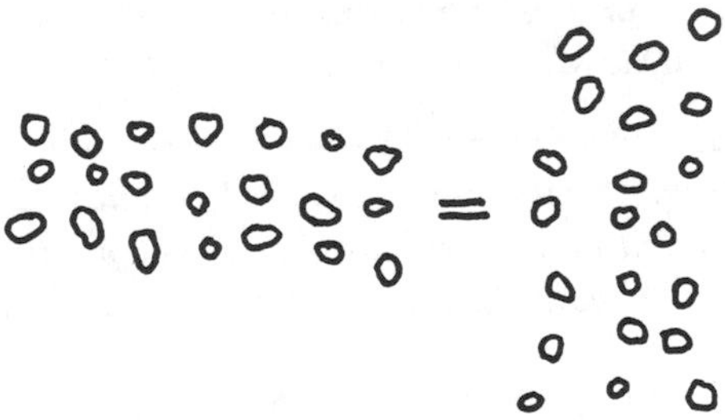

"Three rows of hand-plus-two and hand-plus-two short rows of three
Are just the same according to which way you look, you see!
In brief, a triple heptad is the same as seven trebles,
And may quite possibly be true of other things than pebbles."

Mugg viewed it from all angles, then he gave a raucous belch
And trod on a Batrachian that perished with a squelch.
He growled, "I do not understand these arithmetic quirks,
But maybe we should try it to discover if it works."

So home they went to get their wives and drag them by the hair,
For Mugg had feet-plus-hand-plus-four, while Ogg had just a pair;
But what with all their screeching and their running every way,
At first they would not form a neat rectangular array.

So Ogg he then positioned each by holding of her down
While Mugg with mighty club in hand, just dinted in her crown;
And when they had them all in place, like pebbles, they could see
That three times hand-plus-two in wives was hand-plus-two times three!

Then Ogg he roared in high delight, cartwheeling to and fro
(Carts had not been invented, but he did it, just to show!);
And Mugg he grinned a shaggy grin and slapped a hairy thigh
And said, "It's true, as sure as Pterodactyls learned to fly!"

And then they feasted on their wives in unuxorious zest,
Except for one whose skull was rather thicker than the rest,
And she was sent to dig a pit and bury every bone,
While Mugg and Ogg went off to find a flat unsullied stone.

Then Ogg he sharpened up a flint and scratched upon the rock:
First Arithmetic Theorem—by Ogg the son of Mok.
He drew his little diagrams, and proved, with QED,
That Three times hand-plus-two of x is hand-plus-two times three.

But Mugg the Missing Link grew bored, and left him there alone,
Still scratching with his silly flint upon his silly stone;
And belching, plunged back in the woods on feet toes simple fives,
In search of wives and mammoth-meat, particularly wives!

ANSWERS

THE ONLY SOLUTION to the Venusian problem is 12 + 12 = 101 in a ternary (base-3) notation, therefore a Venusian has three fingers on each hand. (The Venusian sum is equivalent to 5 + 5 = 10 in our decimal notation.) Raymond DeMers wrote to say that if Venusians had three fingers on each hand they would be more likely to use a base-6 notation. He prefers to believe that Venusians have a total of three fingers, one on one hand, two on the other.

Cameron D. Anderson, of Windsor, Ontario, Canada, and Grenville Turner, of Sheffield University, in England, wrote to say that perhaps the Venusian symbols were for a multiplication problem rather than addition. With this interpretation there is an infinite number of solutions. Readers may enjoy proving that the solution with the smallest base is 13 × 13 = 171 in a base-8 notation.

The second question assumed that the ten extended fingers of both hands represented the ten 1's of a binary number equivalent to the decimal $2^{10} - 1$, or 1,023, and asked for a simple method of using the fingers to subtract from such a number a given smaller number n. The answer, supplied by Frederik Pohl in his article cited earlier, is simply to express n as a binary number, using the fingers in the manner explained. Now bend down every extended finger and extend every bent-down finger—the equivalent of changing every 1 to 0 and every 0 to 1. The new number is the desired binary answer.

CHAPTER 9

Möbius Bands

A burleycue dancer, a pip
Named Virginia, could peel in a zip;
But she read science fiction
and died of constriction
Attempting a Möbius strip.
—CYRIL KORNBLUTH

A SHEET OF PAPER has two sides and a single edge that runs all the way around it in the form of a closed curve. Can a sheet of paper have a single edge and only *one* side, so that an ant can crawl from any spot on the paper to any other spot without ever crossing an edge? It is hard to believe, but apparently no one noticed the existence of such one-sided surfaces until a band with a half-twist was described by August Ferdinand Möbius, a German mathematician and astronomer who died in 1868 (in his *Werke*, Vol. 2, 1858). Since then the Möbius strip, as this surface came to be called, has become the best known of the many toys of topology, that flourishing branch of modern mathematics concerned with properties that remain invariant when a structure is given "continuous deformation."

The deformation that preserves topological properties, such as the one-sidedness of Möbius strips, is often explained by asking the reader to imagine that a structure is made of soft rubber that can be molded into any desired shape provided it is not punctured or a part of it removed and stuck back at some other spot. This is a common misconception. The kind of deformation that preserves topological properties must be defined in a much more technical way, involving continuous mapping from point to point. It is quite possible for two structures to be topologically equivalent ("homeomorphic," as topologists like to say) even though one cannot in our three-dimensional space transform one to the other by deforming it like a rubber sheet. A simple example is provided by two rubber Möbius bands that are mirror images of each other because they are twisted in opposite directions. It is impossible to deform one to the other by stretching and twisting, and yet they are topologically identical. The same is true of a Möbius strip and a strip with three or any other odd number of half-twists. All strips with odd half-twists, and their mirror images, are homeomorphic even though none can be changed to another by rubber-sheet deformation. The same is true of all strips (and their mirror images) with *even* half-twists. Such strips are topologically distinct from those with odd half-twists but all of them are homeomorphic with one another [*see Figure 39*].

More strictly, they are homeomorphic in what topologists call an *intrinsic* sense, that is, a sense that considers only the surface itself and not the space in which it may be embedded. It is be-

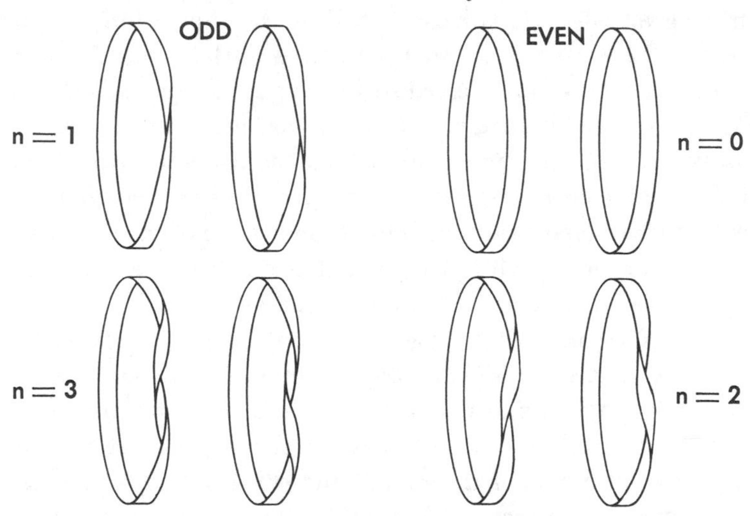

FIGURE 39
Strips with odd (left) and even (right) numbers of half-twists

cause our model of a Möbius strip is embedded in 3-space that it cannot be deformed to its mirror image or to a band with three half-twists. If we could put a paper Möbius band into 4-space, it would be possible to deform it and drop it back into 3-space as a band with any odd number of half-twists of either handedness. Similarly, a band with no twists (topologically equivalent to a cylinder or to a sheet of paper with a hole in it) could be taken into 4-space, twisted, and dropped back into our space with any even number of half-twists of either handedness.

Instead of thinking of the strips as manipulated in 4-space, think of them as zero-thick surfaces in 3-space capable of self-intersection. With a little imagination it is easy to see how a twisted band can be changed, by passing it *through itself*, to any of its topologically equivalent forms. For instance, a "ghost" Möbius strip can be passed through itself to form its mirror image or any surface with an odd number of twists of either handedness.

When a twisted strip is embedded in 3-space, the strip acquires *extrinsic* topological properties it does not have when it is considered apart from its embedding space. Only in this extrinsic sense can one say a Möbius band is topologically distinct from, say, a strip with three half-twists.

The most whimsical intrinsic topological property of the Möbius strip (or any of its intrinsically identical forms) is that when it is cut in half along a line down its middle, the result is not two bands, as one might expect, but a single larger band. As an anonymous limerick has it:

A mathematician confided
That a Möbius strip is one-sided.
You'll get quite a laugh
If you cut it in half,
For it stays in one piece when divided.

Surprisingly, the new band produced by this "bisection" is two-sided and two-edged. Because the model is embedded in 3-space it will have $2n + 2$ half-twists, where n is the number of odd half-twists in the original band. If n is 1, the new strip has four half-twists, an even number, so that it is intrinsically homeomorphic with a cylinder. If n is 3, the final strip has eight half-twists and is tied in a simple overhand knot.

A band with an even number of half-twists (0, 2, 4, . . .) always produces two separate bands when it is cut down the middle, each identical with the original except for being narrower. In 3-space each has n half-twists and the two bands are linked $n/2$ times. Thus when n is 2, bisection produces two bands, each with two half-twists, and they are linked together like two links of a chain. If n is 4, one band is looped twice around the other. When n is 2 you can cut the band to make two linked rings, break one and toss it aside, cut the remaining one to get two still thinner linked rings, break one and toss it aside, and continue (in theory) as long as you like.

In my Dover paperback *Mathematics, Magic, and Mystery*, I explain how magicians have exploited these properties in an old cloth-tearing trick called "the Afghan bands." Stephen Barr suggests another novel way of demonstrating the same properties. He paints a center line around a large, heavy paper strip with a strong water solution of potassium nitrate, then hangs the band on a nail so that only half of the band's width is supported by the nail. When the painted line is touched at the bottom with the burning end of a cigarette, the line burns quickly upward on both sides until the flames meet at the top, then half of the strip drops to produce either one large strip, two linked strips, or a knotted strip depending on whether the original was given one, two, or three half-twists.

Another unexpected result occurs when a strip with odd half-twists is "trisected," that is, if the cut is begun a third of the way from one edge. The cutting takes you twice around the band before you return to the starting point. The result is a band identical with the original except for being narrower (it is the central third of the original), linked with a second band, twice as long, that is identical with (but narrower than) the band that would have been produced by cutting the original in half. When n is 1 (the Möbius strip), trisection produces a small Möbius band linked to a longer two-sided band with four half-twists [*see Figure 40*].

A fascinating puzzle (proposed independently by two readers, Elmer L. Munger and Steven R. Woodbury) now presents itself. After producing the two interlocked strips by trisecting a Möbius band, see if you can manipulate them until they nest together to form the triple-thick Möbius band shown in Figure 40. If you succeed, you will find a curious structure in which two outer "strips" are separated all the way around by a Mö-

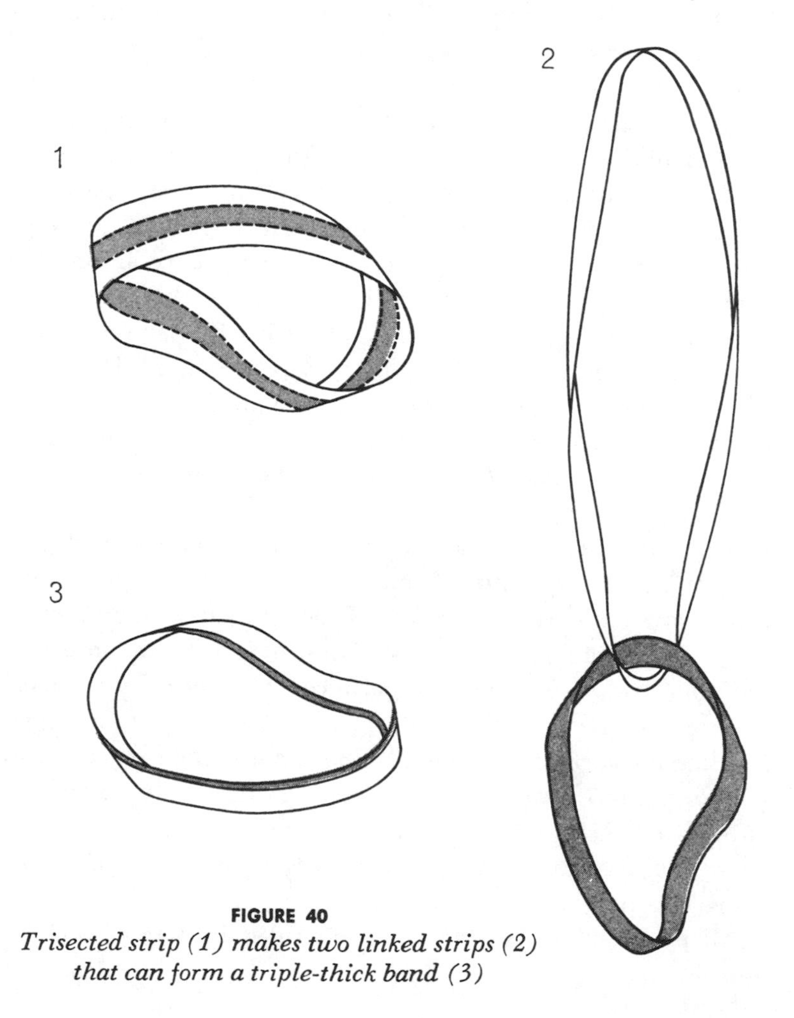

FIGURE 40
Trisected strip (1) makes two linked strips (2) that can form a triple-thick band (3)

bius strip "between" them. One would suppose, therefore, that the Möbius strip is surrounded by two separate bands, but of course you know this is not the case. The same structure can be made by putting three identical strips together, holding them as one, giving them a half-twist, and then joining the three pairs of corresponding edges. If this triple-thick band is painted red on its "outside," you will find it possible to interchange the outside parts so that the red-painted side of the larger band goes into the interior and the triple-thick band is colorless on its "outside." It is amusing to form such bands, of m thicknesses and n half-twists, and then to work out formulas for what results when such bands are bisected and trisected.

The Möbius strip has many strange intrinsic properties. It is what topologists call "nonorientable." Imagine the strip to be a true surface of zero thickness. Embedded in this 2-space are flat creatures that are mirror asymmetric (not identical with their mirror images). If such a creature moves once around the band to rejoin his fellows, he will have changed his parity and will have become a mirror reflection of his former self. (Cosmologists have devised analogous models of twisted 3-space in which it would be possible for an astronaut to make a circuit around the cosmos and return with his heart on the other side.) Remember, you must assume that the flat creatures are "in" the zero-thick surface, not "on" it.

All nonorientable surfaces must contain at least one Möbius surface. Stated differently, from any nonorientable surface one can always cut a Möbius surface. Topologists have found many strange kinds of nonorientable surfaces, such as the Klein bottle, the projective plane, and Boy's surface (discovered by the German mathematician Werner Boy), all of them closed and edgeless like the surface of a sphere. The Klein bottle can be cut in half to produce two Möbius strips, as explained in Chapter 2 of my *Sixth Book of Mathematical Games from Scientific American*. The projective plane becomes a Möbius surface if a hole is made in it.

All nonorientable surfaces are one-sided in 3-space and all

orientable surfaces (in which flat asymmetric creatures are unable to reverse their handedness) are two-sided in 3-space. Sidedness is not an intrinsic topological property like orientability. It is only in our space that we can speak of a two-dimensional surface as having one or two sides, just as we can only speak of a closed one-dimensional line as having an outside and inside when it is embedded in a plane.

Another intrinsic property of the Möbius strip has to do with graph theory. On the plane, or on any band with even half-twists, four is the largest number of points that can be joined by nonintersecting lines that connect every pair of points [*see Figure 41*]. It is not hard to prove that this cannot be done with five points. On a Möbius surface, however, a maximum of six points can be mutually joined by nonintersecting lines. Consider six spots on an opened-out strip [*see Figure 41*]. Assume that the top and bottom ends of this strip are joined after the strip is given one half-twist or any odd number of half-twists. Can you connect every pair of points with a line without having two lines intersect or without cheating by passing a line through a spot? Here again, assume that the strip has no thickness. Each line must be thought of as "in" the paper like an ink line that has soaked through to the other side.

The Möbius strip has practical uses. In 1923 Lee De Forest obtained a United States patent for a Möbius filmstrip that records sound on both "sides." More recently the same idea has been applied to tape recorders so that the twisted tape runs twice as long as it would otherwise. Several patents have been granted for Möbius strip conveyor belts designed to wear equally on both sides. In 1949 O. H. Harris obtained patent No. 2,479,929 on a Möbius abrasive belt. B. F. Goodrich Company secured a similar patent (No. 2,784,834) in 1957. In 1963 J. W. Jacobs obtained patent No. 3,302,795 on a self-cleaning filter belt for dry-cleaning machines. It makes possible easy washing of dirt from both "sides" as the twisted belt goes around.

In 1963 Richard L. Davis, a Sandia Corporation physicist in Albuquerque, invented a Möbius strip nonreactive resistor.

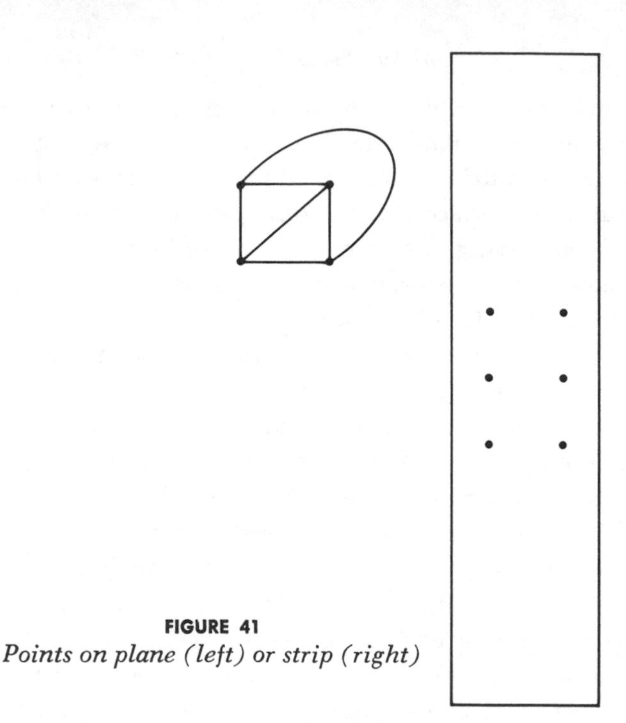

FIGURE 41
Points on plane (left) or strip (right)

Bonding metal foil to both sides of a nonconductive ribbon and then forming a triple-thick Möbius band, Davis found that when electric pulses flowed in both directions around the foil (passing through themselves), the strip acquired all kinds of desirable electronic properties. (See *Time*, September 25, 1964, and *Electronics Illustrated*, November 1969, pages 76 f.)

Modern sculptors have based numerous abstract works on the Möbius surface. The Smithsonian Institution's new Museum of History and Technology in Washington, D.C., has an eight-foot-high steel Möbius band that rotates slowly on a pedestal in front of the entrance. Max Bill, a Swiss sculptor, has based dozens of abstract works on the Möbius strip. [*See Figure 42.*]

Graphic artists have used the strip in advertising and works of art. Two uses of the Möbius band by the Dutch artist Maurits C. Escher are reproduced in Figures 43 and 44. In 1967 Brazil honored a mathematical congress by issuing a commemorative stamp bearing a picture of the Möbius strip. In 1969 a Belgian stamp featured a Möbius band flattened to a triangle. (Both

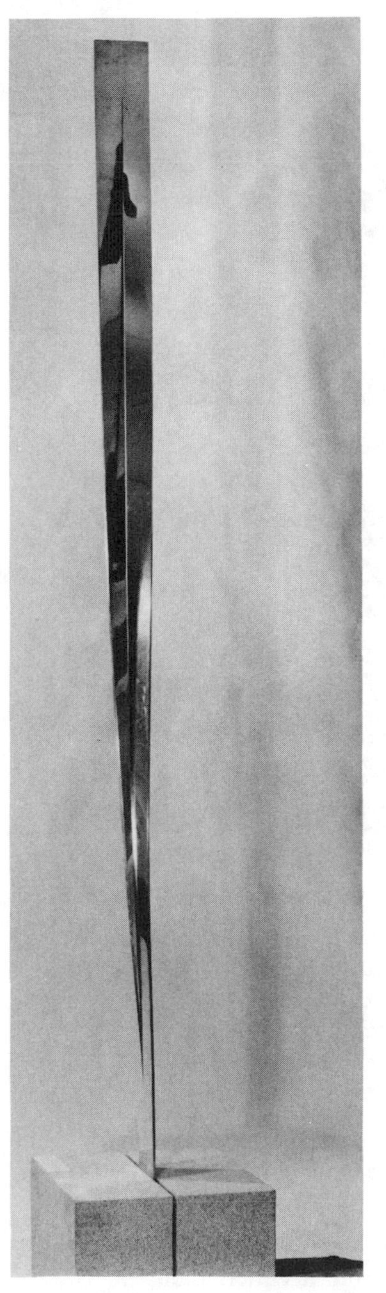

FIGURE 42
Continuous Surface
in Form of a Column (*1953*),
Albright-Knox Art Gallery, Buffalo

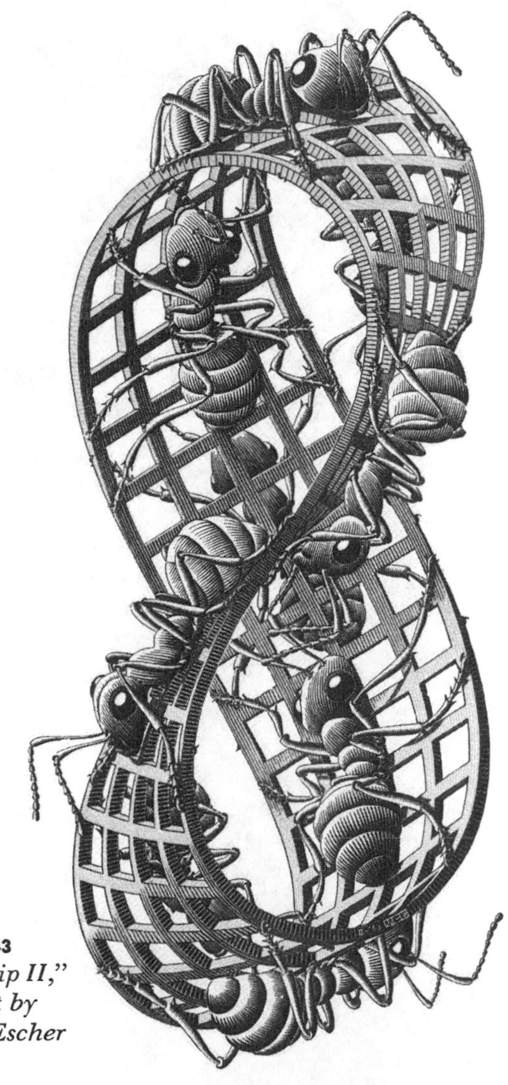

FIGURE 43
"Möbius Strip II,"
a woodcut by
Maurits C. Escher

FIGURE 44
"Möbius Strip I," a 1961 wood engraving by Maurits C. Escher, shows a bisected Möbius band. It is a single band in the form of three fish, each biting the tail of the fish in front.

stamps are shown in Figure 45.) A triangular, flattened Möbius strip was the official symbol of Expo '74, the 1974 World Exposition in Spokane, Washington. *The New Yorker*'s cover, April 5, 1976, showed a Möbius band around which about thirty businessmen are walking in both directions.

The Möbius surface has played a central role in numerous science-fiction stories, from my "No-Sided Professor" to Arthur C. Clarke's "The Wall of Darkness" (*Super Science Stories*, July 1949). Many friends have sent me Christmas cards with messages, such as "Endless joy," printed on Möbius strips. It is curious that if you keep pulling such a printed strip through your fingers, the endless message is always right side up although at any spot on the strip the printing at the back is upside down. When I was an editor of *Humpty Dumpty's Magazine*, I once based an activity feature ("Watch the Thanksgiving Day Parade," November 1955, pages 82–84) on this property.

Writers who type rapidly and are annoyed by having to put new sheets into the machine have been known to adopt the expedient of typing on paper that comes in rolls, like paper towels. If they used a long loop of paper, it could be twisted to permit typing on both sides. Waldo R. Tobler once suggested mapping the world onto a Möbius strip so that the edge coincided with the poles, and the lines of latitude and longitude were symmetrically spaced. If it were done properly, you could puncture the map at any spot and the point on the other "side" would be the spherical antipode.

Hexaflexagons have an odd number of half-twists and are therefore Möbius surfaces. For an introduction to the interesting

FIGURE 45
Brazilian and Belgian stamps that display Möbius bands

topology of "crossed" Möbius bands, see Problem 15 of the next chapter. On the problem of the minimum-length strip that can be folded and joined to make a Möbius surface, see my *Sixth Book of Mathematical Games from Scientific American*, Chapter 6. Acrobatic skiers now perform a trick called the "Möbius flip" in which they make a twist while somersaulting through the air.

A group of French writers and mathematicians who publish outlandish experiments in French word play under the group name of OuLiPo, use Möbius strips for transforming poems. For example, a quatrain on one side of a strip may have the rhyme scheme *abab*, and a quatrain on the other side may rhyme *cdcd*. Twisted into a Möbius band, a new poem, *acbd-acbd* is created. (See the two chapters on the OuLiPo in my *Penrose Tiles to Trapdoor Ciphers* (Freeman, 1989).)

In recent years even nonmathematical writers seem to have become enamored of the Möbius surface as a symbol of endlessness. There is a poem by Charles Olson called "The Moebius Strip," and Carol Bergé's *A Couple Called Moebius: Eleven Sensual Short Stories* (Bobbs-Merrill, 1972) displays a huge Möbius strip on the jacket and smaller strips at the top of each story. "When a man and woman join as lovers," says the book's flyleaf, "there is a potential infinity of relationships that, like the Moebius strip, has no beginning and no end: only a continuum. . . . There is wisdom and honesty in these stories: enough to make one feel a kinship with these people—an acquaintance possibly formed somewhere along the Moebius strip of one's own life."

It is hard to see how a twist in an endless loop adds anything to the metaphor that a simple untwisted band or an old-fashioned circle wouldn't provide. All the twist does is keep bringing one back to previously visited spots in alternate left- and right-handed forms, but how this applies to the filmstrip of one's life is not clear.

The first story, more accurately the start of a story, in John Barth's *Lost in the Funhouse* (Doubleday, 1968) is designed to be read on an actual Möbius surface. The reader is told to cut the page along the dotted line, then twist and paste to make a Möbius band on which he can read the endless "Frame-Tale": "Once Upon A Time There Was A Story That Began Once Upon A Time There Was A Story That Began . . ."

This is the old children's tale that has an infinite beginning with no middle or end. Once I wrote the following middleless metapoem that has an infinite beginning and an infinite end:

One day
A mad metapoet
With little to say
Wrote a mad metapoem
That started:
"One day
A mad metapoet
With little to say
Wrote a mad metapoem
That started:
' "One day

.

.

.

Sort of close," '
Were the words that the poet
Finally chose
To bring his mad poem
To some
Sort of close,"
Were the words that the poet
Finally chose
To bring his mad poem
To some
Sort of close.

Unfortunately I have not yet found an appropriate topological surface on which to print it.

ANSWERS

ONE WAY to solve the Möbius strip puzzle is shown in Figure 46. Assume that the strip is given a half-twist before the ends are joined; points *a, b, c, d, e* at the bottom will then meet corresponding points at the top. The surface must be thought of as having zero thickness, its lines being "in" the strip, not "on" it.

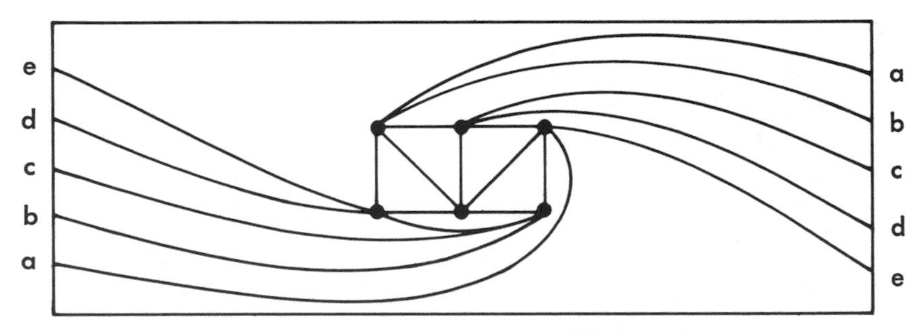

FIGURE 46
Answer to Möbius strip problem

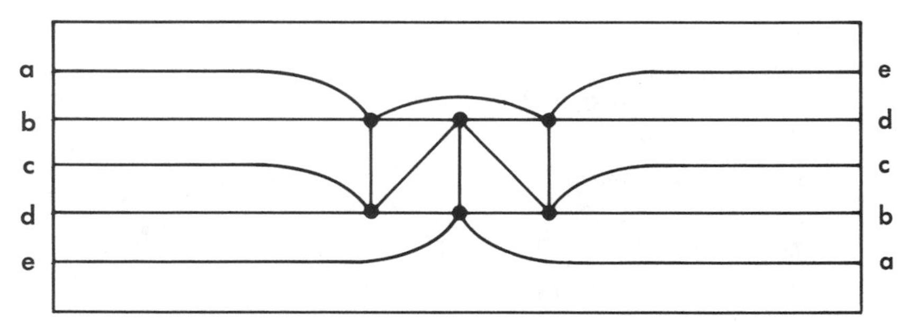

FIGURE 47

The complement of this graph is a map that requires six colors if each region differs in color from each of its neighbors. Figure 47 is another symmetrical solution that was sent by many readers.

CHAPTER 10

Ridiculous Questions

None of the following short problems requires a knowledge of advanced mathematics. Most of them have unexpected or "catch" answers, and are not intended to be taken seriously.

1. In a certain African village there live 800 women. Three percent of them are wearing one earring. Of the other 97 percent, half are wearing two earrings, half are wearing none. How many earrings all together are being worn by the women?

2. Each face of a convex polyhedron can serve as a base when the solid is placed on a horizontal plane. The center of gravity of a regular polyhedron is at the center, therefore it is stable on any face. Irregular polyhedrons are easily constructed that are unstable on certain faces; that is, when placed on a table with an unstable face as the base, they topple over. Is it possible to make a model of an irregular convex polyhedron that is unstable on *every* face?

3. What is the missing number in the following sequence: 10, 11, 12, 13, 14, 15, 16, 17, 20, 22, 24, 31, 100, —, 10000. (Hint: The missing number is in ternary notation.)

4. Among the assertions made in this problem there are three errors. What are they?

a) $2 + 2 = 4$
b) $4 \div \frac{1}{2} = 2$
c) $3\frac{1}{5} \times 3\frac{1}{8} = 10$
d) $7 - (-4) = 11$
e) $-10(6 - 6) = -10$

5. A logician with some time to kill in a small town decided to get a haircut. The town had only two barbers, each with his own shop. The logician glanced into one shop and saw that it was extremely untidy. The barber needed a shave, his clothes were unkempt, his hair was badly cut. The other shop was extremely neat. The barber was freshly shaved and spotlessly dressed, his hair neatly trimmed. The logician returned to the first shop for his haircut. Why?

6. A single cell is added to a ticktacktoe board [*see Figure 48*]. If the game is played in the usual manner, the first player easily obtains three in a row by playing first as shown. If there were no extra cell, the second player could stop a win only by taking the center. But now the first player can move as shown, winning obviously on his next move.

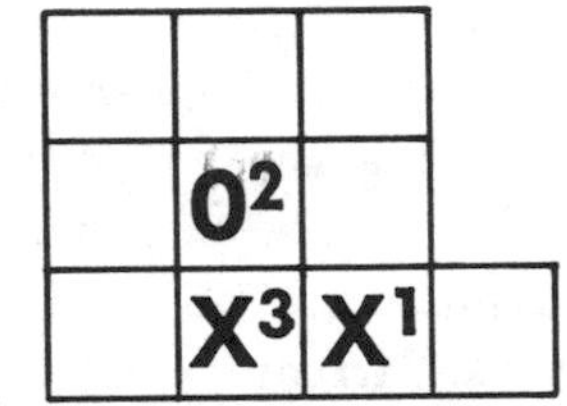

FIGURE 48
A first-player win

Let us modify the game by a new proviso. A player can win on the bottom row only by taking all *four* cells. Can the first player still force a win?

7. A secretary types four letters to four people and addresses the four envelopes. If she inserts the letters at random, each in a different envelope, what is the probability that exactly three letters will go into the right envelopes?

8. Consider these three points: the center of a regular tetrahedron and any two of its corner points. The three points are

coplanar (lie on the same plane). Is this also true of all *ir*regular tetrahedrons?

9. Solve the crossword puzzle in Figure 49 with the help of these clues:

HORIZONTAL

1. Norman Mailer has two.
2. Is indebted.
3. Chicago vehicles.
4. Relaxation.

VERTICAL

1. Skin blemish.
5. Works in the dark.
6. Character in *Wind in the Willows*.
7. Famous white Dixieland trombonist.

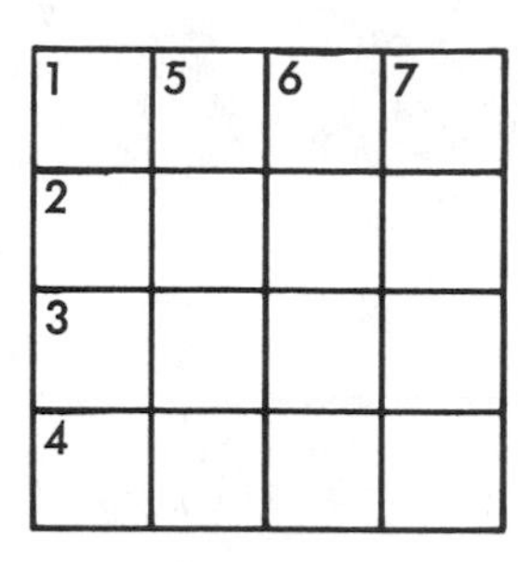

FIGURE 49
Crossword puzzle

10. Three points are selected at random on a sphere's surface. What is the probability that all three lie on the same hemisphere? It is assumed that the great circle, bordering a hemisphere, is part of the hemisphere.

11. If you took three apples from a basket that held 13 apples, how many apples would you have?

12. Two tangents to a circle are drawn [*see Figure 50*] from a point, C. Tangent line segments YC and XC are necessarily equal. Each has a length of 10 units. Point P, on the circle's circumference, is randomly chosen between X and Y. Line AB is then drawn tangent to the circle at P. What length is the perimeter of triangle ABC?

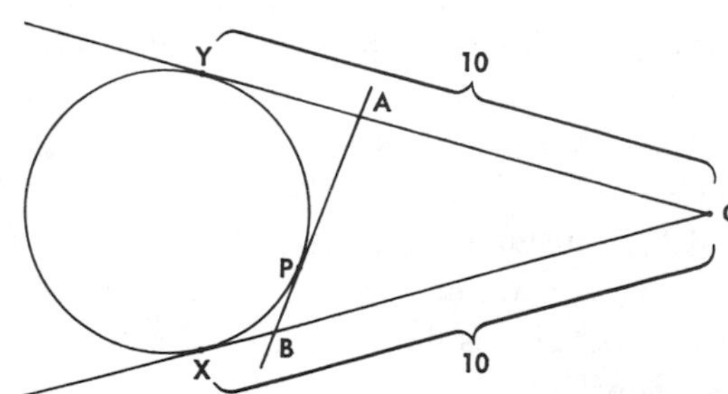

FIGURE 50
A tangent problem

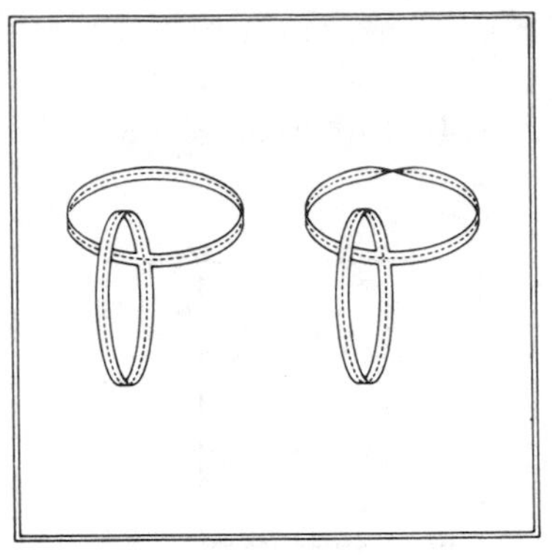

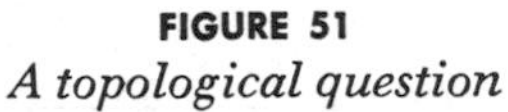

FIGURE 51
A topological question

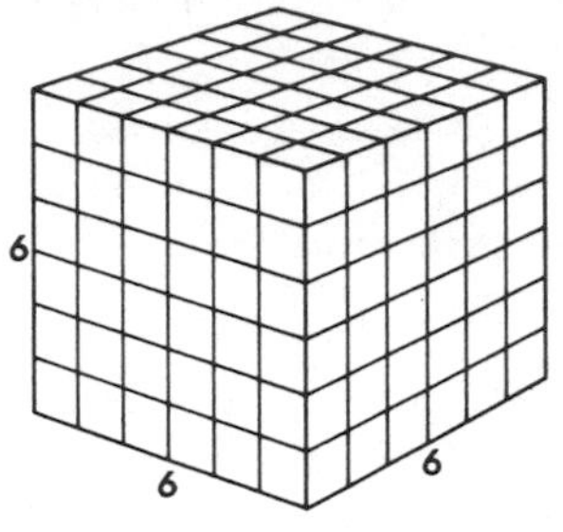

FIGURE 52
Cube-dissection puzzle

13. If nine thousand nine hundred and nine dollars is properly written \$9,909, how should twelve thousand twelve hundred and twelve dollars be written?

14. A chemist discovered that a certain chemical reaction took 80 minutes when he wore a jacket. When he was not wearing a jacket, the same reaction always took an hour and 20 minutes. Can you explain?

15. Each of the two paper structures shown in Figure 51 consists of a horizontal band attached to a vertical band of the same length and width. The structures are identical except that the second has a half-twist in its vertical band. If the first is cut along the broken lines, the surprising result is the large square band shown as a border of the illustration.

What results if the second structure is similarly cut along the broken lines?

16. An equilateral triangle and a regular hexagon have perimeters of the same length. If the triangle has an area of two square units, what is the area of the hexagon?

17. Can a $6 \times 6 \times 6$ cube be made with 27 bricks that are each $1 \times 2 \times 4$ units [*see Figure* 52]?

18. A customer in a restaurant found a dead fly in his coffee. He sent the waiter back for a fresh cup. After taking one sip he shouted, "This is the *same* cup of coffee I had before!" How did he know?

19. A metal sheet has the shape of a two-foot square with semicircles on opposite sides [*see Figure 53*]. If a disk with a diameter of two feet is removed from the center as shown, what is the area of the remaining metal?

20. "I guarantee," said the pet-shop salesman, "that this parrot will repeat every word it hears." A customer bought the parrot but found it would not speak a single word. Nevertheless, the salesman told the truth. Explain.

21. From one corner of a square extend two lines that exactly trisect the square's area [*see Figure 54*]. Into what ratios do these trisecting lines cut the two sides of the square?

22. A 10-foot piece of cylindrical iron pipe has an interior diameter of four inches. If a steel sphere three inches in diameter is inserted into the pipe at end *A* and a steel sphere two inches in diameter is inserted at end *B*, is it possible, with the help of a rod, to push each sphere through the entire length of pipe so that it emerges at the other end?

23. Give at least three ways a barometer can be used to determine the height of a tall building.

24. How can you make a cube with five paper matches? No bending or splitting of matches is allowed.

25. Which situation is more likely after four bridge hands have been dealt: you and your partner hold all the clubs or you and your partner have no clubs?

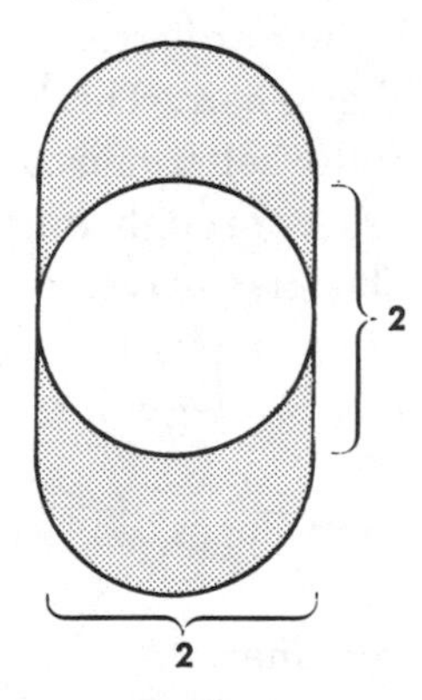

FIGURE 53
Hole-in-metal problem

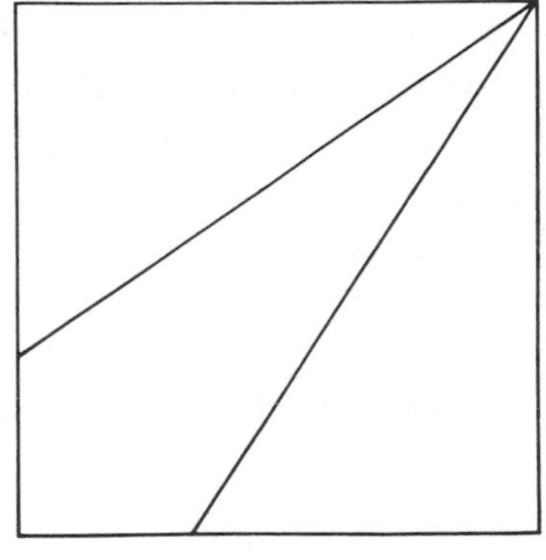

FIGURE 54
Trisecting the square

26. This old-timer still confuses almost everyone who hears it for the first time. Smith gave a hotel clerk $15 for his room for the night. When the clerk discovered that he had overcharged by $5, he sent a bellboy to Smith's room with five $1 bills. The dishonest bellboy gave only three to Smith, keeping the other two for himself. Smith has now paid $12 for his room. The bellboy has acquired $2. This accounts for $14. Where is the missing dollar?

ANSWERS

1. Among the 97 percent of the women, if half wear two earrings and half none, this is the same as if each wore one. Assuming, then, that each of the 800 women is wearing one earring, there are 800 earrings in all.

2. No. If a convex polyhedron were unstable on every face, a perpetual motion machine could be built. Each time the solid toppled over to a new base it would be unstable and would topple over again.

3. Each number is 16 in a number system with a different base, starting with base-16 and continuing with bases in descending order, ending with base-2. The missing number, 16 in the ternary system, is 121.

4. Only equations *b* and *e* are false, therefore the statement that there are three errors is false. This is the third error.

5. Each barber must have cut the other's hair. The logician picked the barber who had given his rival the better haircut.

6. Assume the cells are numbered from 1 through 10, taking them left to right and top to bottom. The first player can win only by taking either cell 2 or 6 on his first move. I leave it to the reader to work out the first player's strategy for all responses.

7. Zero. If three letters match the envelopes, so will the fourth.

8. Yes. *Any* three points in space are coplanar.

9. The solution to the crossword puzzle is shown in Figure 55.

FIGURE 55
Answer to crossword puzzle

M	M	M	M
O	O	O	O
L	L	L	L
E	E	E	E

10. The probability is certainty. Any three points on a sphere are on a hemisphere.

11. Three apples.

12. The triangle's perimeter is 20 units. Lines tangent to a circle from an exterior point are equal, therefore $YA = AP$ and $BP = XB$. Since $AP + BP$ is a side of triangle ABC, it is easy to see that the triangle's perimeter is $10 + 10 = 20$. This is one of those curious problems that can be solved in a different way on the assumption that they have answers. Since P can be anywhere on the circle from X to Y, we move P to a limit (either Y or X). In both cases one side of triangle ABC shrinks to zero as side AB expands to 10, producing a degenerate straight-line "triangle" with sides of 10, 10, and 0 and a perimeter of 20. (Thanks to Philip C. Smith, Jr.)

13. \$13,212.

14. Eighty minutes is the same as one hour and 20 minutes.

15. Cutting the second structure has the same result as cutting the first. Indeed, the result is the same large square regardless of the number of twists in the vertical band! For an additional surprise, see what happens when the untwisted band of the second structure is bisected and the twisted band is *trisected*.

16. Three square units [*see Figure 56*].

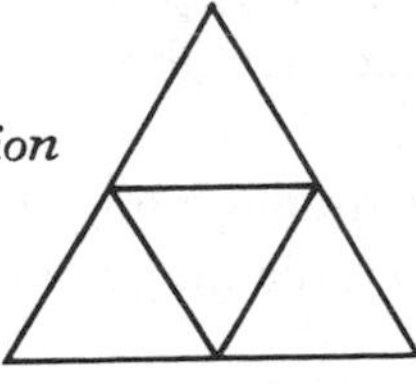

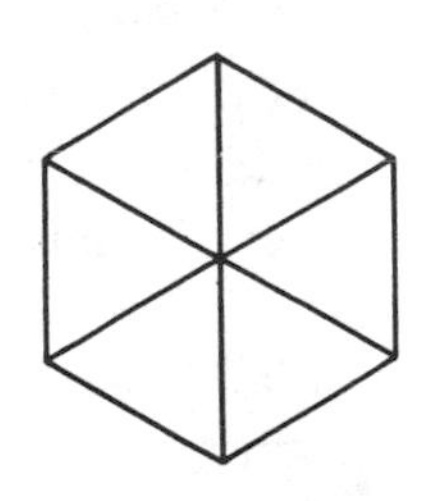

FIGURE 56
Triangle-hexagon solution

17. No. Think of the order-6 cube as made up of 27 cubes with sides of two units and alternately colored black and white. Since 27 is an odd number there will be 13 cubes of one color and 14 of another. No matter how a brick is placed within this cube, half of its unit cubes will be black and half white, so that if the cube can be formed, it must contain as many black unit cubes as white. This contradicts the fact that the large cube has more unit cubes of one color than of the other, therefore there is no way to build the order-6 cube with the 27 bricks.

18. The customer had sugared his coffee before he found the dead fly.

19. The two semicircles together form a circle that fits the hole. The remaining metal therefore has a total area of four square feet.

20. The parrot was deaf.

21. The trisecting lines also trisect each side of the square. As Piet Hein, who sent me this problem, points out, this is easily seen by dividing any rectangle into halves by drawing the main diagonal from the corner where the trisecting lines originate. Each half of the rectangle obviously must be divided by a trisecting line into two triangles such that the smaller is half the area of the larger. Since the two triangles share a common altitude, this is done by making the base of the smaller triangle half the base of the larger.

22. Yes, if the two spheres are pushed through the tube at different times.

23. Here are five:

(1) Lower the barometer by a string from the roof to the street, pull it up, and measure the string.

(2) Same as above except instead of pulling the barometer up, let it swing like a pendulum and calculate the length of the string by the pendulum's frequency. (Thanks to Dick Akers for this one.)

(3) Drop the barometer off the roof, note the time it takes to fall, and compute the distance from the formula for falling bodies.

(4) On a sunny day, find the ratio of the barometer's height to the length of its shadow and apply this ratio to the length of the building's shadow.

(5) Find the superintendent and offer to give him the barometer if he will tell you the height of the building.

Solution (1) is quite old (I heard it from my father when I was a boy), but the most complete discussion of the problem, with all solutions except (2), is in Alexander Calandra's book *The Teaching of Elementary Science and Mathematics* (Ballwin, Mo.: ACCE Reporter, 1969). Calandra's earlier discussion of the problem, in the teacher's edition of *Current Science*, was the basis of a *New York Times* story, March 8, 1964, page 56.

24. If "cube" is taken in the numerical sense, the five matches can be used to form 1 or 27, or VIII, or I with an exponent of 3. If the bottoms of the matches are straight, the arrangement shown in Figure 57 produces a tiny cube at the center.

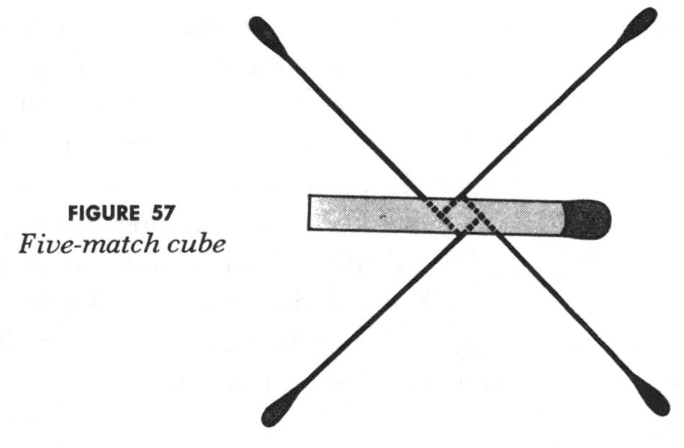

FIGURE 57
Five-match cube

25. The probabilities are the same. If you and your partner have no clubs, all the clubs will have been dealt to the other two players.

26. Adding the bellboy's $2 to the $12 Smith paid for his room produces a meaningless sum. Smith is out $12, of which the clerk has $10 and the bellboy $2. Smith got back $3, which, added to the $12 held by the clerk and the bellboy, accounts for the full amount of $15.

CHAPTER 11

Polyhexes and Polyaboloes

THE USUAL jigsaw puzzle is almost devoid of mathematical interest: the pieces are fitted together by trial and error, and if one has enough determination and patience, the pattern is eventually completed. But if the pieces have simple polygonal shapes, the task of fitting them into a predetermined figure becomes one of combinatorial geometry, offering scope for considerable mathematical ingenuity and sometimes raising questions that are not mathematically trivial. If a set of polygonal pieces is obtained by applying a simple combinatorial rule, it takes on a quality of elegance, and the task of exploring the set's combinatorial properties can be as fascinating as it is time-devouring.

Among recreational mathematics enthusiasts the most popular of such sophisticated jigsaws are the polyominoes. These are pieces formed by joining n unit squares in all possible ways. Many articles have been devoted to them, and Solomon W. Golomb, professor of engineering and mathematics at the University of Southern California, has written a book about them, *Polyominoes*. By joining equilateral triangles along their edges one obtains another well-explored family of shapes known as polyiamonds. The hexiamonds (polyiamonds formed with six equilateral triangles) are discussed in my *Sixth Book of Mathematical Games from Scientific American.*

Many readers who enjoy working with polyominoes and polyiamonds have written to propose other ways of obtaining a basic set of polygons that can be used for similar recreations. In this chapter I shall discuss the two sets that have prompted the most correspondence. Very little has been published about either of them.

Since there are only three regular polygons that tile the plane —squares, equilateral triangles, and regular hexagons—one thinks at once of forming pieces by joining congruent hexagons. There is only one way to join two hexagons; there are three ways to join three hexagons and seven ways to join four of them. Because these shapes resemble the structural diagrams of benzene-ring compounds, two readers, Eleanor Schwartz and Gerald J. Cloutier, have suggested calling them "benzenes." Other names have been proposed, but it seems to me that the best is "polyhexes," the name adopted by David Klarner, who was one of the first to investigate them. The 7 tetrahexes, with names culled from the letters of many different readers, are shown in Figure 58. The next-largest set, the pentahex, has 22

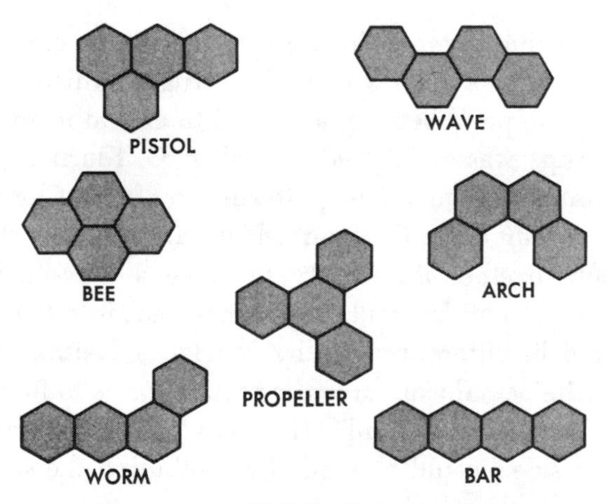

FIGURE 58
The seven distinct tetrahexes

distinct shapes—a bit too unwieldy for recreational uses. There are 82 hexahexes, 333 heptahexes, and 1,448 octahexes. (Because pieces may be turned over, mirror-image forms are not considered distinct.) As with polyominoes and polyiamonds, no formula is known by which the number of polyhexes of a given order can be determined.

The reader is urged to cut a set of tetrahexes from cardboard. (If you have a hexagonally tiled floor, you could make the pieces correspond in cell size to the tiles so that the floor can be used as a background for working on tetrahex problems.) Of the eight symmetrical patterns in Figure 59, all but one can be formed from a full set of the seven tetrahexes. Many readers proposed the "rhombus," the "triangle," and the "tower." The "ink blot" and the "grapes" are from Richard A. Horvitz, the "annulus" from Cloutier, the "pyramid" from Klarner, and the "rug" from both T. Marlow and Klarner. Can you identify the impossible figure? No simple proof of its impossibility has yet been found. (The tower is not the impossible one, although it is difficult and has a unique solution except for a trivial reversal of two pieces that together form a mirror-symmetrical shape.) All seven pieces must be used, and solutions obtainable by reflection of the entire figure are not, of course, counted as different.

Many striking symmetrical shapes can be formed with the 22 pentahexes [*see Figure 60*]. The "rug" (which can be bisected, the two parts being placed end to end to form a longer, narrower rug) was discovered by Robert G. Klarner (father of David Klarner). The other patterns are from Christoph M. Hoffman of Hamburg, Germany. Note that the two rhombuses can be put together differently to make a 5-by-22 rhombus. (This bisected rhombus and the rhombus cut into two triangles were solved in different ways by Marlow.) Neither the tetrahexes nor the pentahexes have the required area to form a hexagon, but both Marlow and Miss Schwartz discovered that a hexagon of side 4 could be made by combining the seven tetrahexes with the three trihexes.

Turning to polygonal units that are not regular, we find that

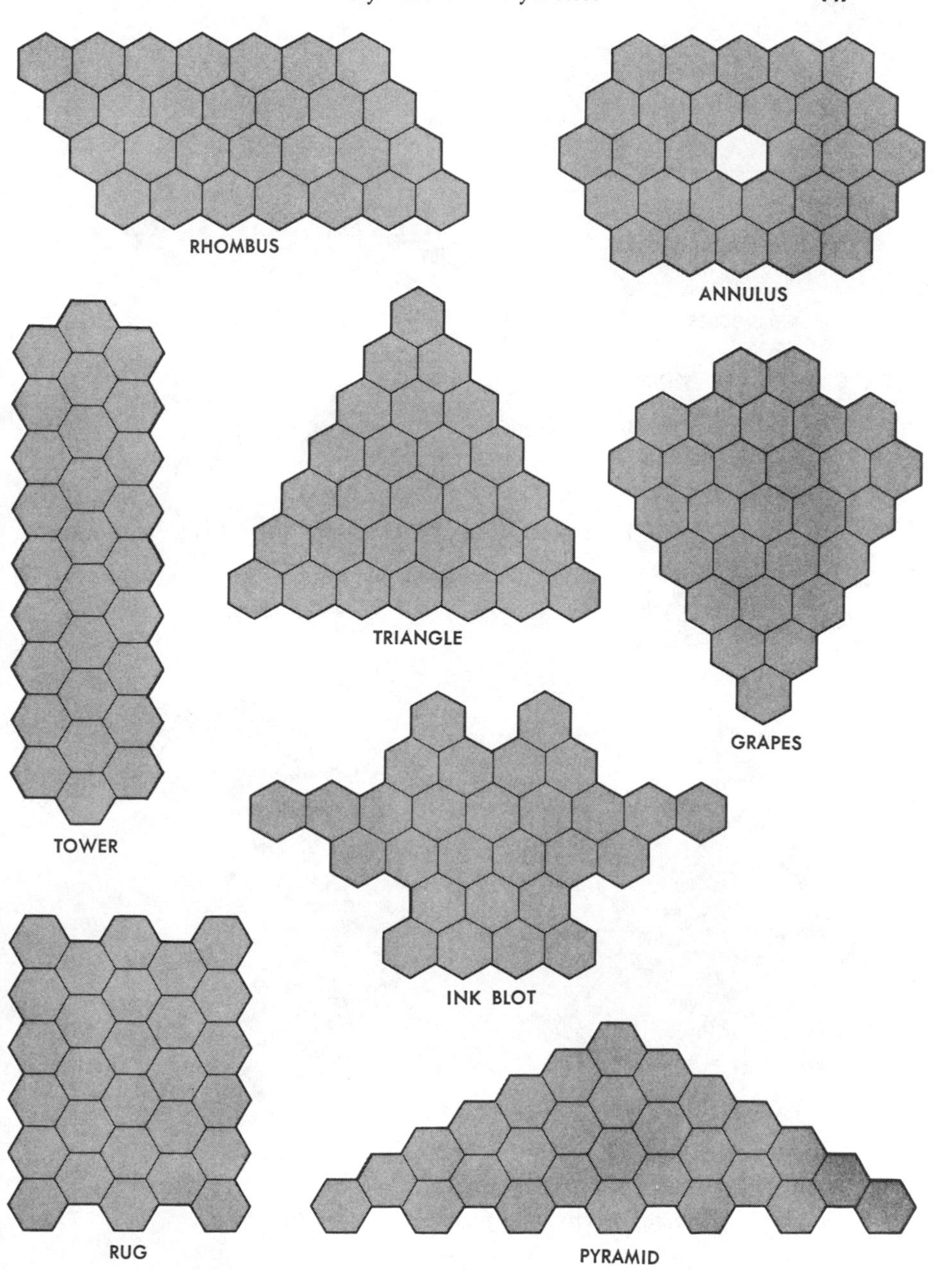

FIGURE 59

Patterns to be formed with tetrahexes, all but one of them possible

FIGURE 60

Some patterns to be formed with pentahexes

the simplest are the isosceles right triangles. They can be joined either at their sides or along their hypotenuses. We shall speak of the sides as *s* edges and the hypotenuses as *h* edges. This family of pieces was first discussed in print by Thomas H. O'Beirne in *New Scientist*, December 21, 1961, in a column on recreational mathematics that he then contributed regularly to the magazine. The pieces had been suggested to him by S. J. Collins of Bristol, England, who gave the name "tetraboloes" to the order-4 set because the diabolo, a juggling toy, has two isosceles right triangles in its cross section. This implies the generic name "polyaboloes." There are 3 diaboloes, 4 triaboloes, 14 tetraboloes, 30 pentaboloes, and 107 hexaboloes.

The set of 14 tetraboloes in Figure 61 has a total area of 28 *s*-unit squares or 14 *h*-unit squares. Since neither 14 nor 28 is a square number there is no possibility of forming a square with the complete set. A *2s*-by-*2s* square has the right area to be formed with two tetraboloes, but this proves to be impossible. There are three squares that can be formed with subsets of the

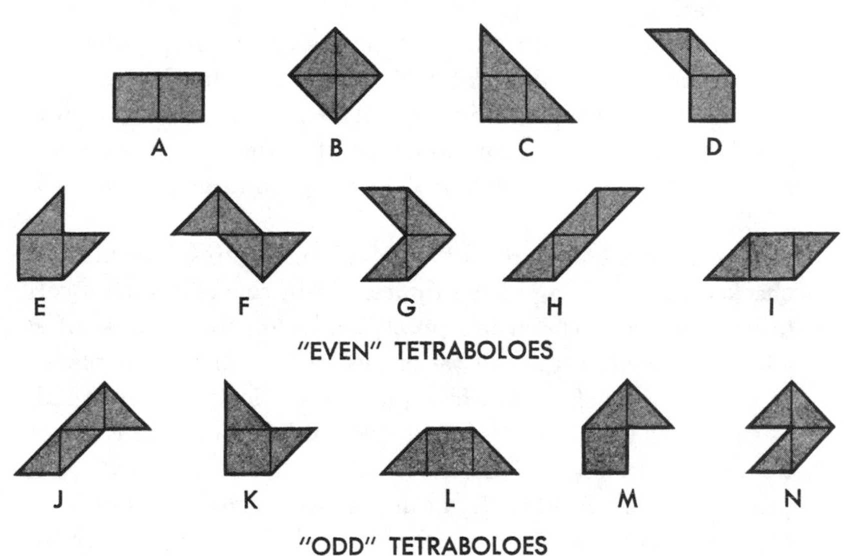

FIGURE 61
The 14 distinct tetraboloes

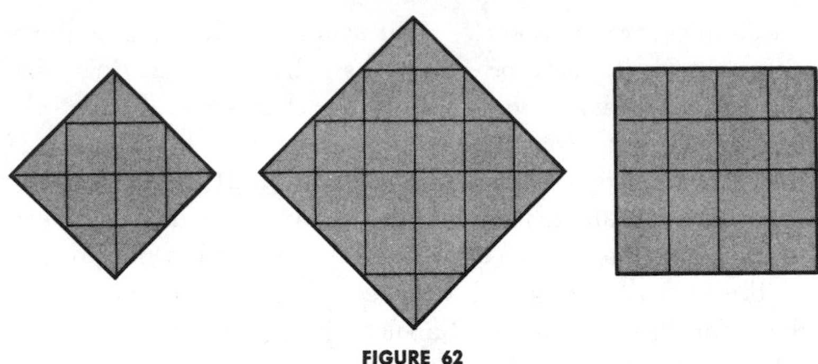

FIGURE 62
Squares with tetrabolo solutions

complete set [*see Figure 62*]. If the reader will make a set of cardboard tetraboloes, he will find that it can be a pleasant task to discover patterns for these three squares. The smallest square has only two solutions and the number of solutions for the two larger ones is not known.

Figure 63 shows all the rectangles with s-edge sides that have the proper area to be formed with the full set or a subset of the 14 pieces; Figure 64 shows all such rectangles with h-edge sides. (Rectangles with a side of $1h$ are clearly impossible except, of course, for the $1h$-by-$1h$.) Note that the largest-area rectangle of each type, calling for all 14 pieces, is said to be impossible. I shall give a remarkable proof of this that was discovered by O'Beirne and explained in his column of January 18, 1962.

Most impossibility proofs for polyomino figures depend on a checkerboard coloring of the figure, but in this case such a coloring is no help. O'Beirne's proof focuses on the number of h edges possessed by the full set of pieces. If each piece is placed so that the sides of its unit triangles are vertical and horizontal, as in Figure 61, its h edges will slope either up to the right or up to the left. Piece A has no h edges. It and the next eight pieces (B, C, D, E, F, G, H, I) are called "even" pieces because each of them has an even number of h edges sloping in each direction. (Zero is considered an even number.) The last five pieces (J, K, L, M, N) are "odd" because each has an odd number of h edges sloping in each direction. Because there is an odd

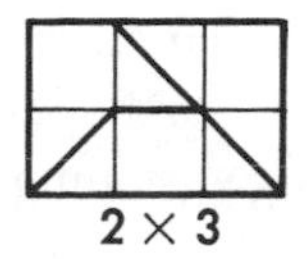

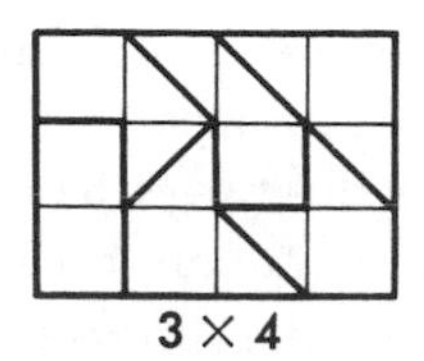

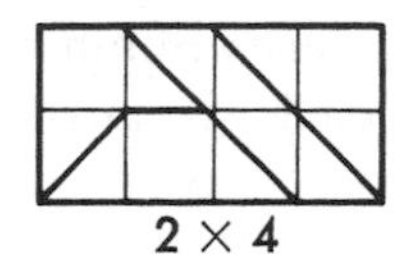

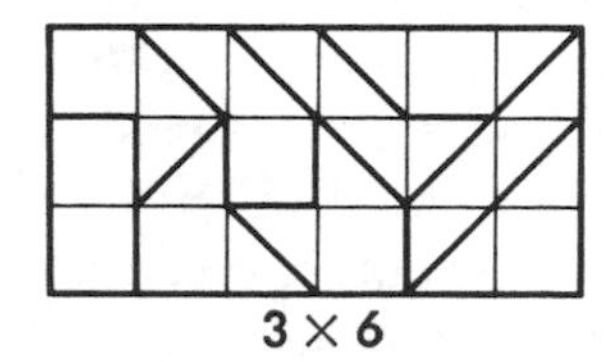

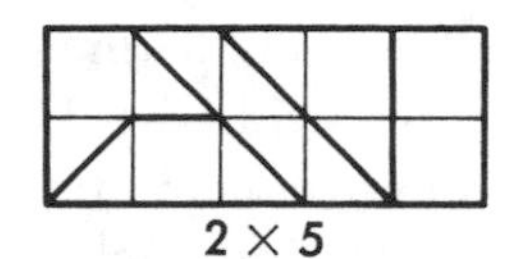

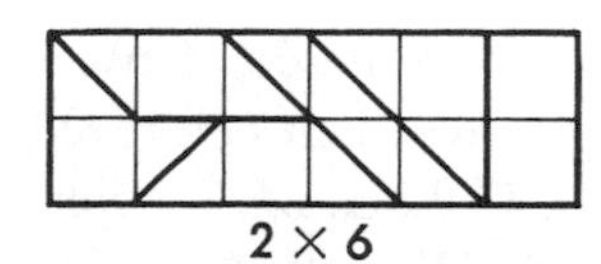

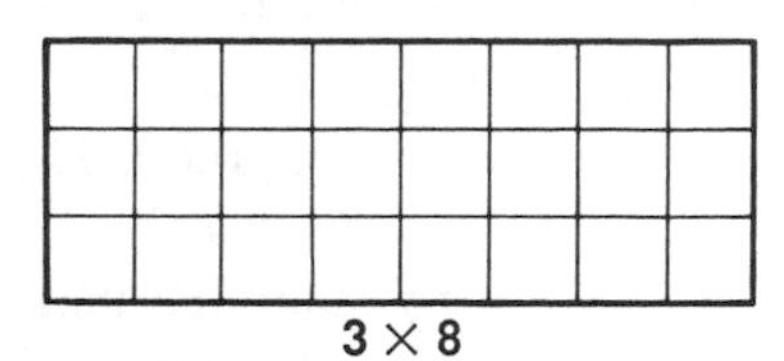

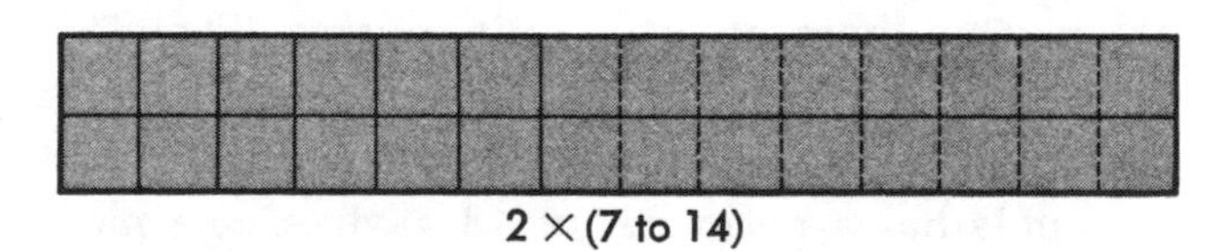

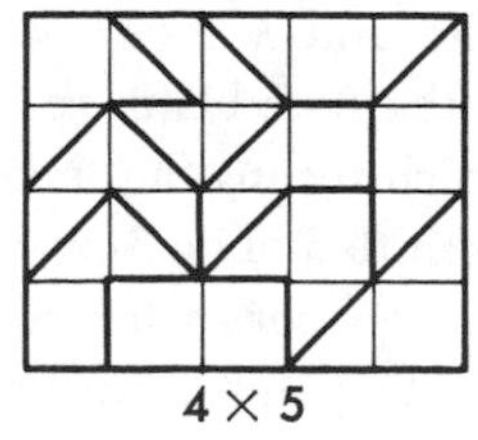

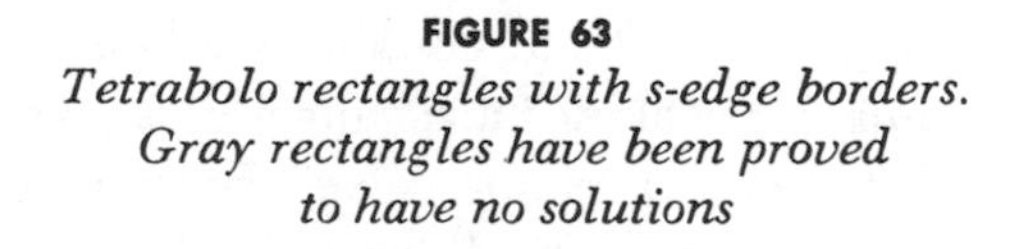

FIGURE 63
Tetrabolo rectangles with s-edge borders. Gray rectangles have been proved to have no solutions

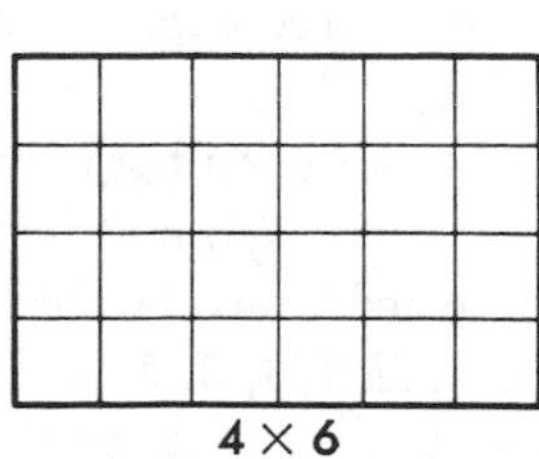

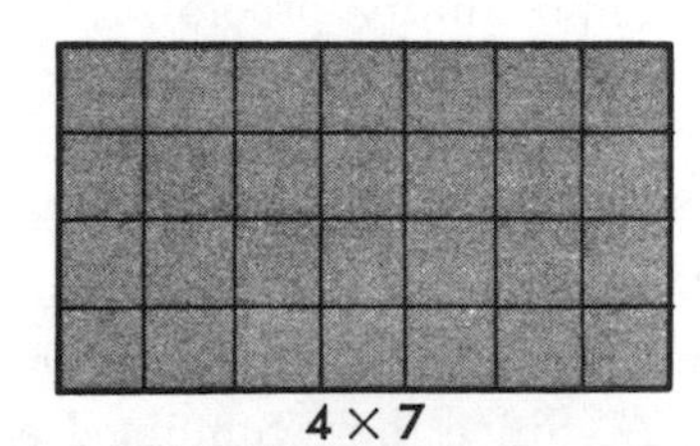

number of odd pieces, it follows that no matter how the entire set is arranged in a pattern with its *s* edges oriented orthogonally, there will always be an odd number of *h* edges sloping one way and an odd number sloping the other way.

Now consider the two rectangles that require the full set of 14 pieces. Clearly on each rectangle there must be an *even* number of *h* edges sloping in each direction. Within each rectangle every *h* edge is paired with another *h* edge sloping the same way and therefore the number of internal *h* edges, of either type, must be even. And on the rectangle that has a perimeter of *h* edges we count an even number of each type along the perimeter. Neither rectangle, therefore, can be made with the 14 pieces. The proof does not apply to all bilaterally symmetric figures. You might like to prove this by constructing such a 14-piece figure.

The gray rectangles in Figure 63 that have a height of two *s* units and a width equal to or greater than seven can be proved impossible by observing that six pieces (*B*, *D*, *E*, *G*, *M*, *N*) cannot be placed within any of them without dividing the field into areas that are not multiples of two *s*-unit squares. Therefore they cannot appear in the pattern. The remaining eight pieces can contribute a maximum of 17 *s* edges to the perimeter. But the 2-by-7 rectangle has a perimeter of 18 *s* edges, one more than this subset can provide.

A sample pattern is shown for all *s*-edge and *h*-edge rectangles for which solutions are known. Are the four blank rectangles possible? Each calls for 12 pieces, which means that one "odd" piece and one "even" piece must be omitted. The 3-by-8 rectangle may be impossible, because its large perimeter severely limits the number of ways pieces can be placed, but no impossibility proofs for any of the four blank rectangles are known, nor have solutions been found.

More ambitious readers may wish to tackle a difficult square pattern proposed by O'Beirne. Discard the six symmetrical shapes that are unaltered when they are turned over and consider only the eight asymmetric shapes: *D*, *F*, *H*, *I*, *J*, *K*, *M*, and *N*. Since their combined area is 16 *s*-unit squares, they might form an order-4 square, but such a square has 16 *s* units in its border whereas the eight pieces can contribute no more than 12 *s* units to the perimeter. Suppose, however, we consider each piece in its two mirror-image forms, making a set of 16 pieces in all. Pieces may not, in this case, be turned over; that is, each "enantiomorphic" pair must be used as a set of two mirror images. The 16 pieces have a total area of 16 *h*-unit squares. Will they form a square with sides of four *h* units? O'Beirne found that they would, but such patterns are extremely hard to come by and none has been published.

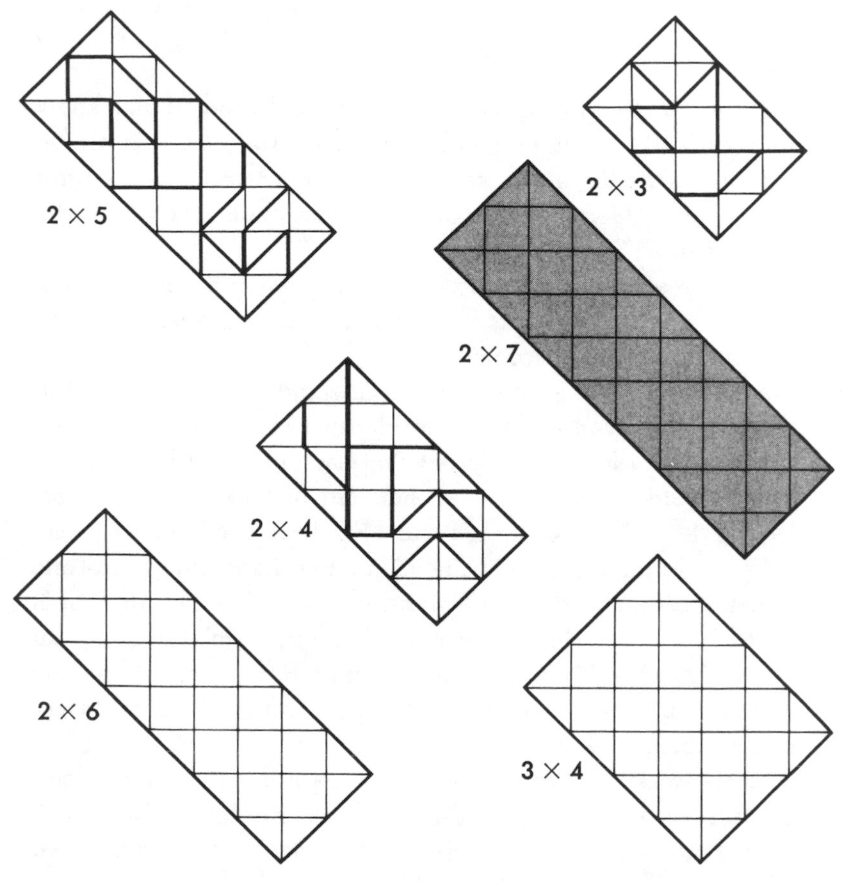

FIGURE 64

Rectangles with h-edge borders. Gray one is impossible

The tetraboloes also provide an answer—perhaps the simplest—to an unusual question asked by C. Dudley Langford of Ayrshire, Scotland, and passed on to me by a British mathematician, H. Martyn Cundy. Langford wanted to know if there are four shapes of equal area, no two of them alike (mirror images not being considered different), that can be put together in four different ways to make four larger replicas of each shape. All four pieces must be used in each replica. I discovered a simple solution with a set of four tetraboloes. Can the reader pick out the four and show how they replicate themselves?

ADDENDUM

SETS OF tetrahexes have been marketed in Europe, but I know of no such sets in the United States. In 1971 a set of ten plastic polyhexes (the three trihexes and the seven tetrahexes), with a booklet of problems by Stewart T. Coffin, was sold under the trade name of Snowflake.

Many readers provided proofs of the uniqueness of the tetrahex tower solution (in its two variants). In all proofs the key was the propeller's placement.

Andrew C. Clarke, of Cheshire, England, reported that each polyhex of orders 4 and 5 would tile the plane, and all but four of the order-6. He also reported that each of the order-4 polyaboloes would tile the plane, all but four of order-5, and all but 19 of order-6. None of these results has been confirmed.

W. F. Lunnon, of the Atlas Computer Laboratory, Chilton, England, enumerated the polyhexes through order-12 in his paper "Counting Hexagonal and Triangular Polyominoes," in *Graph Theory and Computing*, edited by R. C. Read (Academic Press, 1972). The counts for orders 9 through 12 are, respectively, 6,572; 30,490; 143,552; and 683,101.

Little work has been done on enumerating the polyaboloes. Several readers agreed that there are 318 of order-7. Charles W. Trigg counted 1,106 of order-8, and Robert Oliver counted 3,671 of order-9.

ANSWERS

THE IMPOSSIBLE tetrahex figure among those displayed in Figure 59 is the triangle. David Klarner was able to prove it impossible by an argument that begins by observing the limited number of positions in which the propeller can be placed.

In Klarner's solution for the difficult tower figure [*see Figure 65*], note that the shaded portion has an axis of symmetry allowing it to be reflected and thus providing a second solution.

One way to form a square with the eight asymmetric tetraboloes and their mirror images (no turning over being allowed) was independently discovered in 1962 by two Britons, R. A. Setterington of Taunton and A. F. Spinks of Letchworth [*see Figure 66*]. Pieces *G*, *H*, and *M* form a shape that can be rotated and reflected; *JKN* and *FJK* can each be rotated, and *CE* can be interchanged with *MP* to provide many variant solutions. Thomas H. O'Beirne of Glasgow recently found a different way to arrange *A*, *B*, *C*, *D*, *E*, *F*, *G*, and *H* to form a pattern in which rotations and interchanges of parts produce still other variants. The number of distinct solutions is far from known.

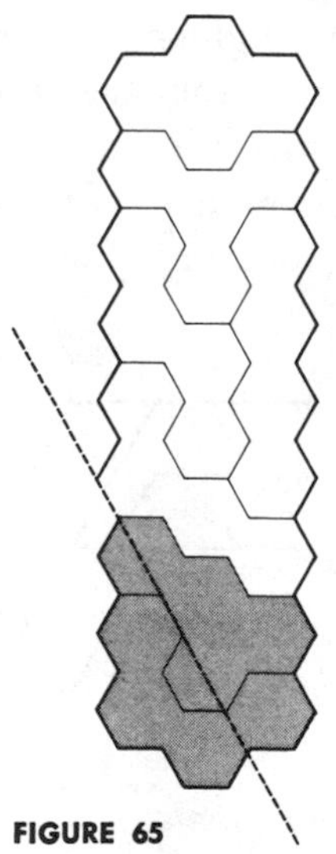

FIGURE 65
The tower tetrahex

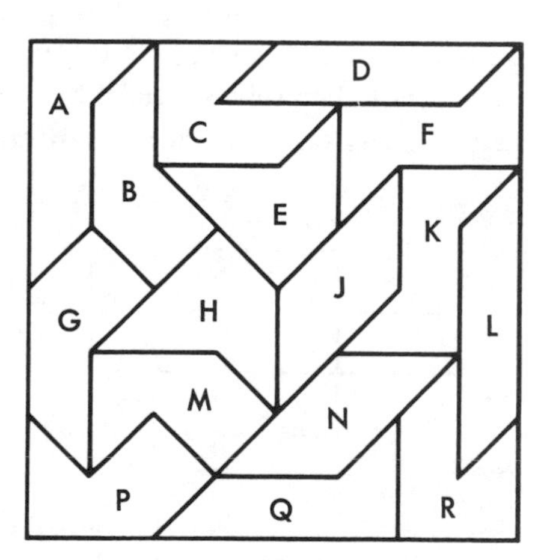

FIGURE 66
Solution of difficult tetrabolo problem

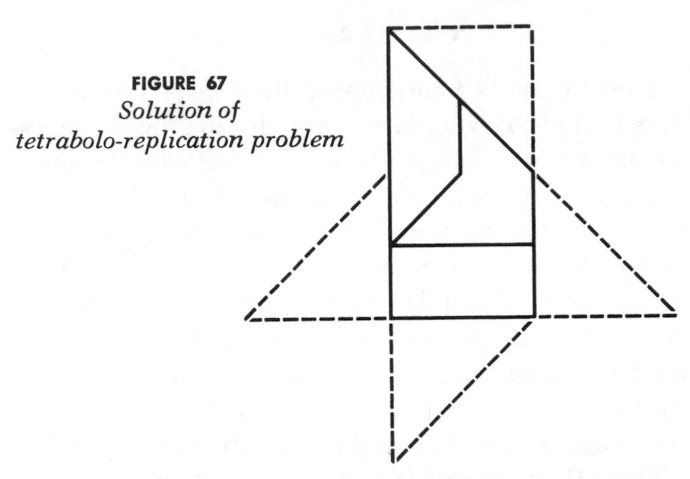

FIGURE 67
Solution of tetrabolo-replication problem

Four tetraboloes can be put together to make larger replicas of each [*see Figure 67*]. Only the triangle need be moved. Note that still other positions of the triangle produce replicas of four more tetraboloes, making the total eight, or more than half of the entire set. Wade E. Philpott and others found that tetraboloes *C, I, K, L* also solve this problem.

Are there other sets of four different shapes with the same property? Maurice J. Povah of Blackburn, England, has proved the number to be infinite. His proof derives from the solution shown at the left in Figure 68, in which four octominoes form replicas of themselves. An affine transformation [*right*], varying the angles, furnishes an infinity of solutions. Povah also

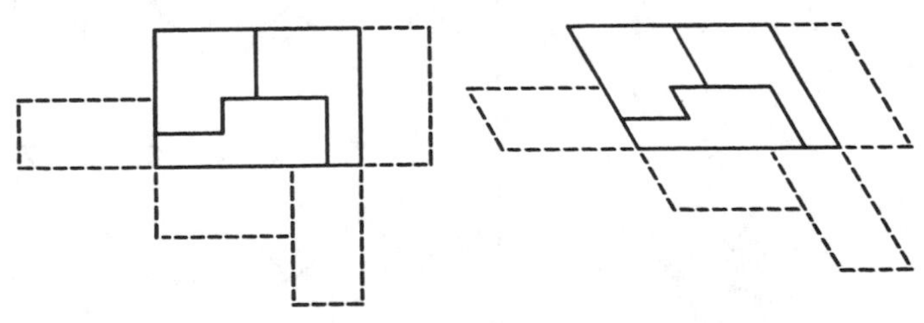

FIGURE 68
Octomino solution with infinite variants

discovered a solution with four hexominoes [*see Figure 69*]. These pieces will replicate 15 different hexominoes, including themselves. Povah believes this is a maximum for the hexominoes. With pentominoes the best he could do with four pieces was to form four replicas of other pentominoes, not including any of their own shapes.

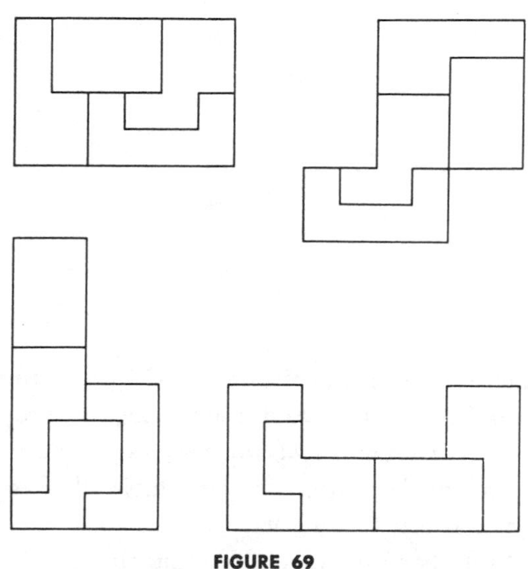

FIGURE 69
Hexomino solution to replication problem

John Harris cleared up the problems of the four unsolved tetrabolo rectangles by proving ingeniously that the 3-by-8 and the 4-by-6, among the rectangles with *s* sides, are impossible, and by supplying solutions for the two *h*-edge rectangles, 2-by-6 and 3-by-4.

CHAPTER 12

Perfect, Amicable, Sociable

ONE WOULD BE hard put to find a set of whole numbers with a more fascinating history and more elegant properties, surrounded by greater depths of mystery—and more totally useless—than the perfect numbers and their close relatives, the amicable (or friendly) numbers.

A perfect number is simply a number that equals the sum of its proper divisors; that is, of all its divisors except itself. The smallest such number is 6, which equals the sum of its three divisors, 1, 2, and 3. The next is 28, the sum of $1 + 2 + 4 + 7 + 14$. Early commentators on the Old Testament, both Jewish and Christian, were much impressed by the perfection of those two numbers. Was not the world created in six days and does not the moon circle the earth in twenty-eight? In *The City of God*, Book 11, Chapter 30, St. Augustine argues that although God could easily have created the world in an instant, He preferred to take six days because the perfection of 6 signifies the perfection of the universe. (Similar views had been advanced earlier by the first-century Jewish philosopher Philo Judaeus in the third chapter of his *Creation of the World.*) "Therefore," St. Augustine concludes, "we must not despise the science of numbers, which, in many passages of Holy Scripture, is found to be of eminent service to the careful interpreter."

The first great achievement in perfect-number theory was Euclid's ingenious proof that the formula $2^{n-1}(2^n - 1)$ always gives an even perfect number if the parenthetical expression is a prime. (It is never a prime unless the exponent n also is prime, although if n is prime, $2^n - 1$ need not be, indeed rarely is, prime.) It was not until 2,000 years later that Leonhard Euler proved that this formula gives *all* even perfects. In what follows, "perfect number" will mean "even perfect number" because no odd perfects are known and they probably do not exist.

To get an intuitive grasp of Euclid's remarkable formula and see how closely it ties perfect numbers to the familiar doubling series, 1, 2, 4, 8, 16 . . . , consider the legendary story about the Persian king who was so delighted with the game of chess that he told its originator he could have any gift he wanted. The man made what seemed to be a modest request: he asked for a single grain of wheat on the first square of the chessboard, two grains on the second square, four on the third, and so on up the powers of 2 to the sixty-fourth square. It turns out that the last square would require 9,223,372,036,854,775,808 grains. The total of all the grains is twice that number minus 1, or a few thousand times the world's annual wheat crop.

In Figure 70 each square of a chessboard is labeled with the number of grains it would hold. Taking one grain from a square, n, leaves $2^n - 1$ grains, the parenthetical expression of Euclid's formula. If this number is a prime, multiply it by the number of grains on the preceding square, the 2^{n-1} of the formula. *Voilà*, we have a perfect number! Primes of the form $2^n - 1$ are now called Mersenne primes after the seventeenth-century French mathematician who studied them. The shaded squares in the illustration mark the cells that become Mersenne primes after losing one grain and that consequently provide the first nine perfect numbers.

From Euclid's formula it is not difficult to prove all kinds of weird and beautiful properties of perfect numbers. For example,

2^0	2^1	2^2	2^3	2^4	2^5	2^6	2^7
2^8	2^9	2^{10}	2^{11}	2^{12}	2^{13}	2^{14}	2^{15}
2^{16}	2^{17}	2^{18}	2^{19}	2^{20}	2^{21}	2^{22}	2^{23}
2^{24}	2^{25}	2^{26}	2^{27}	2^{28}	2^{29}	2^{30}	2^{31}
2^{32}	2^{33}	2^{34}	2^{35}	2^{36}	2^{37}	2^{38}	2^{39}
2^{40}	2^{41}	2^{42}	2^{43}	2^{44}	2^{45}	2^{46}	2^{47}
2^{48}	2^{49}	2^{50}	2^{51}	2^{52}	2^{53}	2^{54}	2^{55}
2^{56}	2^{57}	2^{58}	2^{59}	2^{60}	2^{61}	2^{62}	2^{63}

FIGURE 70

Powers of 2 on a chessboard. Shaded squares yield Mersenne primes

all perfects are triangular. This means that a perfect number of grains can always be arranged to form an equilateral triangle like the ten bowling pins or the fifteen pool balls. Put another way, every perfect number is a partial sum of the series $1 + 2 + 3 + 4 + \ldots$ It is also easy to show that every perfect number except 6 is a partial sum of the series of consecutive odd cubes: $1^3 + 3^3 + 5^3 + \ldots$

The digital root of every perfect number (except 6) is 1. (To obtain the digital root add the digits, then add the digits of the result, and continue until only one digit remains. This is the same as casting out nines. Thus to say that a number has a digital root of 1 is equivalent to saying that the number has a remainder of 1 when divided by 9.) The proof involves showing that Euclid's formula gives a number with a digital root of 1 whenever n is odd, and since all primes except 2 are odd, perfect numbers belong to this class. The one even prime, 2, provides the only perfect number, 6, that does not have 1 as its digital root. Perfect numbers (except 6) can also be shown to be evenly divisible by 4, and equal to 4 (modulo 12).

Because perfect numbers are so intimately related to the powers of 2, one might expect them to have some kind of striking pattern when expressed in the binary system. This proves to be correct. Indeed, given the Euclidean formula for a perfect number, one can instantly write down the number's binary form. Readers are invited first to determine the rule by which this can be done and then to see if they can show that the rule always works.

It is momentarily surprising to learn that the sum of the reciprocals of *all* the divisors (including the number itself) of a perfect number is 2. For example, take the case of 28:

$$\frac{1}{1}+\frac{1}{2}+\frac{1}{4}+\frac{1}{7}+\frac{1}{14}+\frac{1}{28}=2.$$

This theorem follows almost immediately from the definition of a perfect number, n, as the sum of its proper divisors. The sum of all its divisors obviously is $2n$. Let $a, b, c, \ldots$ be all the divisors. We can express the equality as follows:

$$\frac{n}{a}+\frac{n}{b}+\frac{n}{c}+\ldots=2n.$$

Dividing all terms by n produces:

$$\frac{1}{a}+\frac{1}{b}+\frac{1}{c}+\ldots=2.$$

The converse is also true. If the reciprocals of all divisors of n add to 2, n is perfect.

The two greatest unanswered questions about perfects are:

Is there an odd perfect number? Is there an infinity of even perfect numbers? No odd perfect has yet been found, nor has anyone proved that such a number cannot exist. (In 1967 Bryant Tuckerman showed that an odd perfect, if it exists, must be greater than 10^{36}.) The second question hinges, of course, on whether there is an infinity of Mersenne primes, since every such prime immediately leads to a perfect number. When each of the first four Mersenne primes (3, 7, 31, and 127) is substituted for n in the formula $2^n - 1$, the formula gives a higher Mersenne prime. For more than seventy years mathematicians hoped this procedure would define an infinite set of Mersenne primes, but the next possibility, $n = 2^{13} - 1 = 8{,}191$, let them down: in 1953 a computer found that $2^{8{,}191} - 1$ was not a prime. No one knows whether the series of Mersenne primes continues forever or has a highest member.

Øystein Ore, in his *Number Theory and Its History*, quotes a once plausible prediction from Peter Barlow's 1811 book, *Theory of Numbers*. After giving the ninth perfect, Barlow adds that it "is the greatest that will ever be discovered, for, as they are merely curious without being useful, it is not likely that any person will attempt to find one beyond it." In 1876 the French mathematician Edouard Lucas, who wrote a classic four-volume work on recreational mathematics, announced the next perfect to be discovered, $2^{126}(2^{127} - 1)$. The twelfth Mersenne prime, on which it is based, is one less than the number of grains on the last square of a *second* chessboard, if the doubling plan is carried over to another board. Years later Lucas had doubts about this number, but eventually its primality was established. It is the largest Mersenne prime to have been found without the aid of modern computers.

Figure 71 lists the formulas for the twenty-four known perfects, the number of digits in each number, and the numbers themselves until they get too large. The twenty-third perfect came to light in 1963 when a computer at the University of Illinois discovered the twenty-third Mersenne prime. The university's mathematics department was so proud of this that for many years its postage meter stamped the prime on its enve-

	FORMULA	NUMBER	NUMBER OF DIGITS
1	$2^{1}(2^{2}-1)$	6	1
2	$2^{2}(2^{3}-1)$	28	2
3	$2^{4}(2^{5}-1)$	496	3
4	$2^{6}(2^{7}-1)$	8,128	4
5	$2^{12}(2^{13}-1)$	33,550,336	8
6	$2^{16}(2^{17}-1)$	8,589,869,056	10
7	$2^{18}(2^{19}-1)$	137,438,691,328	12
8	$2^{30}(2^{31}-1)$	2,305,843,008,139,952,128	19
9	$2^{60}(2^{61}-1)$		37
10	$2^{88}(2^{89}-1)$		54
11	$2^{106}(2^{107}-1)$		65
12	$2^{126}(2^{127}-1)$		77
13	$2^{520}(2^{521}-1)$		314
14	$2^{606}(2^{607}-1)$		366
15	$2^{1,278}(2^{1,279}-1)$		770
16	$2^{2,202}(2^{2,203}-1)$		1,327
17	$2^{2,280}(2^{2,281}-1)$		1,373
18	$2^{3,216}(2^{3,217}-1)$		1,937
19	$2^{4,252}(2^{4,253}-1)$		2,561
20	$2^{4,422}(2^{4,423}-1)$		2,663
21	$2^{9,688}(2^{9,689}-1)$		5,834
22	$2^{9,940}(2^{9,941}-1)$		5,985
23	$2^{11,212}(2^{11,213}-1)$		6,751
24	$2^{19,936}(2^{19,937}-1)$		12,003

FIGURE 71

The twenty-four known perfect numbers

lopes. [*See top of Figure 72.*] In 1971, at IBM's research center in Yorktown Heights, New York, Tuckerman found the twenty-fourth Mersenne prime. [*See bottom of Figure 72.*] This number, of 6,002 digits, is the largest known prime. It supplied, of course, the twenty-fourth perfect number.

The end digits of perfect numbers present another tantalizing mystery. It is easy to prove from Euclid's formula that an even perfect must end in 6 or 8. (If it ends in 8, the preceding digit is 2; if it ends in 6, the preceding digit must be 1, 3, 5, or 7 except in the cases of 6 and 496.) The ancients knew the first four perfects—6, 28, 496, and 8,128—and from them rashly concluded that the 6's and 8's alternated as the series continued. Scores of mathematicians from ancient times through the Renaissance repeated this dogmatically, without proof, particularly after the fifth perfect number (first correctly given in an anonymous fifteenth-century manuscript) turned out to end in 6. Alas, so does the sixth. The series of terminal digits for the twenty-four known perfects is

6, 8, 6, 8, 6, 6, 8, 8, 6, 6, 8, 8,
6, 8, 8, 8, 6, 6, 6, 8, 6, 6, 6, 6.

There are infuriating hints of order. The first four digits alternate 6 and 8, then 66 and 88 alternate four times. Next, a lone 6 introduces the first triplet of 8's, followed by the first

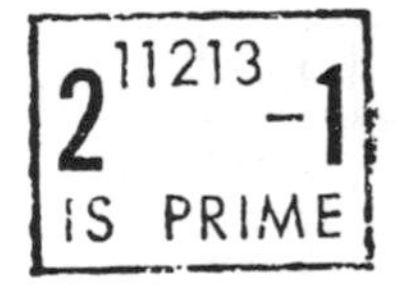

FIGURE 72
Postage-meter stamp and a letterhead honoring the two largest known primes

IBM

Thomas J. Watson Research Center
P.O. Box 218
Yorktown Heights, New York 10598

$2^{19937}-1$ is a prime

triplet of 6's and a lone 8. Finally, the first quadruplet, 6666, appears. Are the digits trying to tell us something? Probably not. Although no one has found a rule for predicting the last digit of the next, undiscovered perfect, it is easy to determine the digit if you know the number's Euclidean formula. Can the reader find a simple rule?

Numbers that are 1 more or 1 less than the sum of their proper divisors have been called "almost perfect." All powers of 2 are almost perfect numbers of the +1 type. No other +1 almost perfects are known, nor have they been proved impossible. No −1 almost perfects have been found, and they too have not yet been proved not to exist.

Amicable numbers derive from an obvious generalization of the perfects. Suppose we start with any number, add its divisors to obtain a second number, then add the divisors of *that* number and continue the chain in the hope of eventually getting back to the original number. If the first step immediately restores the original number, the chain has only one link and the number is perfect. If the chain has two links, the two numbers are said to be amicable. Each is equal to the sum of the divisors of the *other*. The smallest such numbers, 220 and 284, were known to the Pythagoreans. The proper divisors of 220 are 1, 2, 4, 5, 10, 11, 20, 22, 44, 55, and 110. They add to 284. The proper divisors of 284 are 1, 2, 4, 71, and 142. They add to 220.

The Pythagorean brotherhood regarded 220 and 284 as symbols of friendship. Biblical commentators spotted 220 in Genesis 32:14 as the number of goats given Esau by Jacob. A wise choice, the commentators said, because 220, being one of the amicable pair, expressed Jacob's great love for Esau. During the Middle Ages this pair of numbers played a role in horoscope casting, and talismans inscribed with 220 and 284 were believed to promote love. One poor Arab of the eleventh century recorded that he once tested the erotic effect of *eating* something labeled with 284, at the same time having someone else swallow 220, but he failed to add how the experiment worked out.

It was not until 1636 that another pair of amicable numbers, 17,296 and 18,416, were discovered by the great Pierre de Fermat. He and René Descartes independently rediscovered a rule for constructing certain types of amicable pairs—a rule they did not know had previously been given by a ninth-century Arabian astronomer. Using this rule, Descartes found a third pair: 9,363,584 and 9,437,056. In the eighteenth century Euler drew up a list of sixty-four amicable pairs (two of which were later shown to be unfriendly). Adrien Marie Legendre found another pair in 1830. Then in 1867 a sixteen-year-old Italian, B. Nicolò I. Paganini, startled the mathematical world by announcing that 1,184 and 1,210 were friendly. It was the second-lowest pair and had been completely overlooked until then! Although the boy probably found it by trial and error, the discovery put his name permanently into the history of number theory.

More than 1,000 amicable pairs are now known. (Figure 73 lists all pairs smaller than 100,000.) The most complete listing is in a three-part monograph, "The History and Discovery of Amicable Numbers," by Elvin J. Lee and Joseph Madachy

1	220	284
2	1,184	1,210
3	2,620	2,924
4	5,020	5,564
5	6,232	6,368
6	10,744	10,856
7	12,285	14,595
8	17,296	18,416
9	63,020	76,084
10	66,928	66,992
11	67,095	71,145
12	69,615	87,633
13	79,750	88,730

FIGURE 73
Amicable pairs with five or fewer digits

(*Journal of Recreational Mathematics*, Vol. 5, Nos. 2, 3, and 4, 1972). The largest pair on this list of 1,107 pairs has 25 digits each. Discovered too late to be included are several pairs found in 1972 by H. J. J. te Riele, of Amsterdam, the largest of which has 152 digits in each number. So far as I know, this is the largest known pair.

All known amicable pairs have the same parity: two even numbers or (more rarely) two odd numbers. No one has yet proved that a pair of mixed parity is impossible. Every odd amicable pair so far discovered is a multiple of 3. It has been conjectured that this holds for all odd amicables. There is no known formula for generating all amicable pairs, nor is it known whether their number is infinite or finite.

In 1968 I noticed that all even amicable pairs seem to have a sum that is a multiple of 9, and conjectured that this is always the case. Lee shot the conjecture down by finding three counter-examples among the known amicables, and later finding eight more among new pairs of his own discovery. (See "On Division by Nine of the Sums of Even Amicable Pairs," by Elvin J. Lee, *Mathematics of Computation*, Vol. 22, July 1969, pages 545–48.) Because all these counter-instances are numbers with digital roots of 7, I modified my conjecture to: Except for even amicable pairs of numbers equal to 7 (modulo 9), the sums of all pairs of even amicable numbers equal 0 (modulo 9).

If the chain that leads back to the original number has more than two links the number is called "sociable." Until 1969 only two such chains were known. Both were announced in 1918 by P. Poulet, a French mathematician. One is an order-5 chain: 12,496; 14,288; 15,472; 14,536; 14,264. The other is a truly astonishing chain of twenty-eight links that starts with 14,316. This is the largest known chain. (Note that 28 is a perfect number, and that if the 3 in the lowest link is moved to the front you have pi to four decimals.)

Suddenly, in 1969, Henri Cohen, of Paris, discovered eight sociables of order 4. (See his paper "On Amicables and Sociable Numbers," *Mathematics of Computation*, Vol. 24, 1970, pages 423–29). More order-4 sociables were later found by others. The total now stands at fourteen such chains, the smallest numbers of which are listed below:

1,264,460
2,115,324
2,784,580
4,938,136
7,169,104
18,048,976
18,656,380
28,158,165
46,722,700
81,128,632
174,277,820
209,524,210
330,003,580
498,215,416

No sociables higher than order-4 are known except for the order-5 and order-28 chains found in 1918. The biggest unsolved problem is whether a 3-link chain, known as a "crowd," exists. No one has come up with any reason why such chains are impossible, but neither has anyone found an example. Computer sweeps of least numbers up to 60 billion have been made without success. Useless though crowds may be, such searches are likely to continue until a triple chain is encountered or until some clever number theorist proves their impossibility.

ANSWERS

GIVEN A perfect number's Euclidean formula, what simple rule provides the number's binary form? The formula: $2^{n-1}(2^n - 1)$. The rule: Put down n ones followed by $n - 1$ zeros. Example: Perfect number $2^{5-1}(2^5 - 1) = 496$ has the binary form 111110000.

The rule is easily understood. In binary form 2^n is always 1 followed by n zeros. The expression on the left side of Euclid's formula, 2^{n-1}, therefore has the binary form of 1 followed by $n - 1$ zeros. The parenthetical expression $(2^n - 1)$, or one less

than the nth power of 2, has the binary form of n ones. The product of these two binary numbers obviously will be n ones followed by $n - 1$ zeros.

Readers will find it amusing to test the theorem that the sum of the reciprocals of the divisors of any perfect number (including the number itself as a divisor) is 2, by writing the reciprocals in binary form and then adding.

There are several ways to state rules for determining the final digit of a perfect number by inspecting its Euclidean formula, but the following seems the simplest. It applies to all perfect numbers except 6. If the first exponent $(n - 1)$ is a multiple of 4, the perfect number ends in 6. Otherwise it ends in 28.

CHAPTER 13

Polyominoes and Rectification

SOLOMON W. GOLOMB's book *Polyominoes*, published by Scribner's, stimulated worldwide interest in these figures: polygons formed by joining unit squares along their edges. This interest in turn led Golomb, who teaches electrical engineering and mathematics at the University of Southern California, to devote more of his off-duty hours to exploring some of the darker corners of the field. A communication from him deals entirely with a series of fascinating problems, only partially solved, relating to a pentomino game he invented many years ago.

The 12 possible pentominoes (five-square polyominoes) are shown in Figure 74 with Golomb's mnemonic names for them. To play the standard pentomino game you will need these 12 pieces, cut from cardboard, and a standard eight-by-eight checkerboard with squares the same size as the squares that form the pieces. If the reader has never played this game, he is urged to prepare a set of pentominoes and try it; it is one of the most unusual mathematical board games of recent years.

Two players sit across the empty board with the 12 pentom-

FIGURE 74
The 12 pentominoes

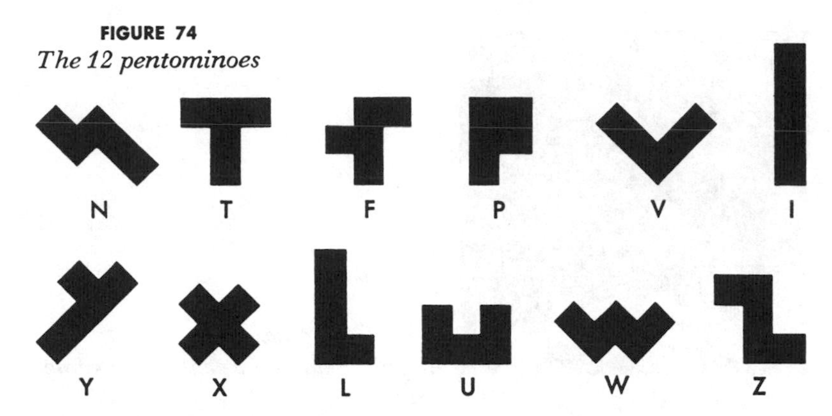

inoes spread on the table beside it. The first player takes one of the pieces and places it so as to cover any five squares of the board. The second player similarly places one of the remaining 11 pieces to cover five of the remaining empty squares. Play alternates until one player is unable to move, either because no remaining piece will fit or because no pieces are left. The player who cannot play is the loser. Games are short, and of course no draw is possible. Nevertheless, great skill and insight are necessary for good play. Mathematicians are far from knowing if the first or the second player can always win if he plays correctly. "The complete analysis of the game," Golomb writes, "is just at the limit of what might be performed by the best high-speed electronic computer, given a generous allotment of computer time and a painstakingly sophisticated program." The most useful strategy, Golomb explains, is to try to split the board into separate and equal areas. There is then an excellent chance that each move by your opponent in one region can be matched by your next move in the other region. If this continues, you are sure to have the last move.

A typical game during which both players keep this strategy in mind has been constructed by Golomb and is shown in Figure 75. Player A puts the *X* near the center to prevent his opponent from splitting the board. Player B counters by fitting the *U* against the *X* (move 2)—a good move, says Golomb, because it "does not simplify the situation for the opponent or al-

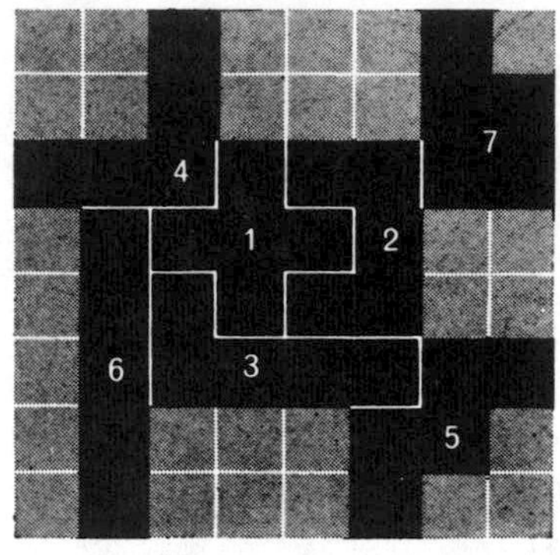

FIGURE 75
A typical standard pentomino game

low him to split the board." Player A is now equally cagey. His *L* move (move 3) continues to prevent a split. The fourth move, by B, is weak because it allows A to place the *W* (move 5) in such a way that it splits the board into two equal regions of 16 squares each. In this case the regions are also identical in shape.

Player B now plays the *I* (move 6), hoping his opponent will not find a piece that fits the other region. But A is able to place the *P* (move 7) in such a way that he wins. Although three of the remaining regions are each large enough to hold a piece, the only three pentominoes that can fit those regions—the *I*, *P*, and *U*—have already been played.

The most interesting variation of the standard game, Golomb continues, is one he calls "choose-up pentominoes." Instead of selecting a piece at each move, players alternate in choosing pieces before the game begins, until each owns six. The last to choose plays first, and the game continues as before except that each player must play only his own pieces. The strategy of this game is quite different from that of the preceding one. Instead of using the split to create a situation in which an even number of moves remains, a player tries to leave as many moves as possible for his own pieces and as few as possible for his opponent's. He does this by creating what Golomb calls "sanctuaries": regions into which only his own pieces will fit.

Golomb's comments on the typical choose-up game shown in Figure 76 are as follows. Player A gets rid of the *X*, his most troublesome piece, by playing it as shown in step 1 in the illustration. (Pieces chosen by A and B are listed beside each board and shown crossed off after use.) Player B places his *W*, another difficult piece (step 2). A now uses his *F* (step 3) to create a sanctuary for his *Y*. B uses his *L* (step 4) to make a sanctuary for his *U*. A plays his *N* (step 5). B places his *I* (step 6) to create a two-by-three rectangle that accommodates only two remaining pieces: his *P* and *U*. A then plays his *V* (step 7) but resigns when he sees that the plays must continue as shown (step 8). The sanctuaries are filled in turn and A is left with a *T* pentomino that will not fit.

If players of this game are unequal in skill, Golomb suggests that the better player give himself a handicap by letting his opponent choose first and play last. A bigger handicap allows the weaker of the two players to make the first two, or even three, choices as well as play last.

There are still other pleasant variations of the game. In "deal-out pentominoes" the names or pictures of the pieces are placed on cards. The deck is shuffled and dealt. Each player takes the pieces indicated by his cards and the game proceeds as in the choose-up form. In "partnership pentominoes" four players sit on four sides of the board and take turns playing, with opposite players forming teams. The team of the first player unable to move is the losing team. Any of the three previously described games can be played in this way. In "cutthroat pentominoes," which also applies to all three games, three or more players participate, but each is on his own. The last to play is the winner. He gets 10 points per game. The first person unable to play scores nothing and all others get 5 points.

Now for a sampling of some new problems suggested to Golomb by the standard game when it is played on square boards of various sizes. The board must be at least 3 by 3 to allow a first move, and of course the first player must win on the 3-by-3 since no second move can be made. On the 4-by-4 board it turns out that the second player can always win. Golomb has set out all possible first plays—excluding reflections and rotations—and a winning reply [*see Figure 77*]. In every case but one, the second player has a choice of winning moves. How long will it

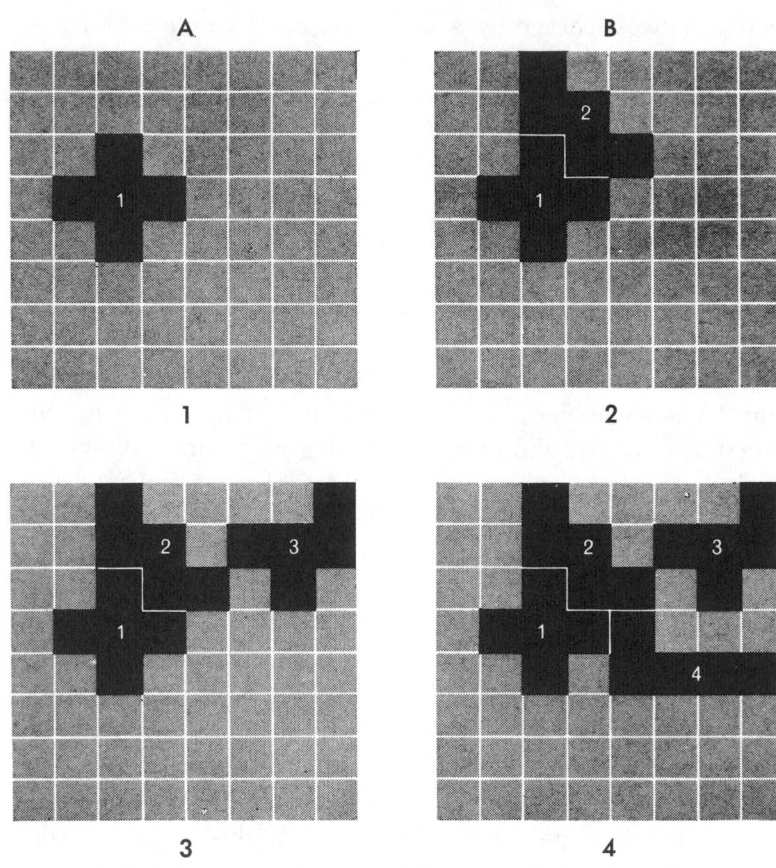

FIGURE 76 *A game of "choose-up pentominoes"*

take the reader to identify the game in which the second player has only the one winning response shown?

It might be supposed that the 5-by-5 board would be more difficult to analyze than the 4-by-4, but surprisingly it is much simpler. The reason is that there is a first move that can easily be shown to lead to a victory for the first player. Can the reader discover it?

The order-6 board brings an enormous jump in complexity; no one yet knows which player has the advantage. "Several promising moves have been subjected to exhaustive analysis and found to allow the second player to win," Golomb writes,

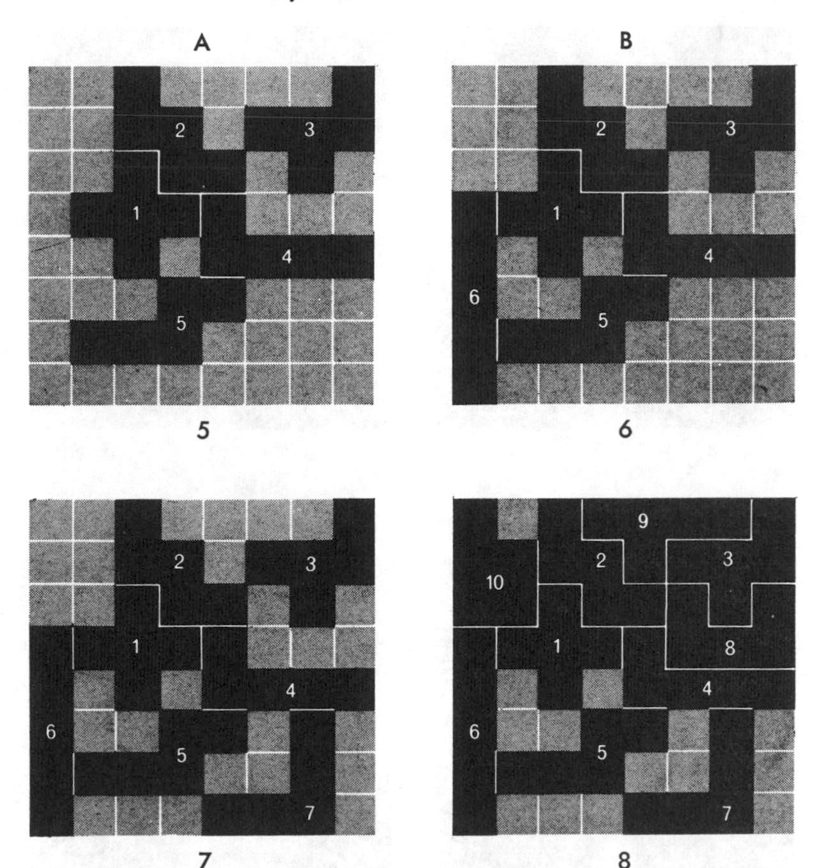

"but the complete analysis will be quite lengthy, involving the correct follow-up strategies against each of a very large number of possible first moves."

Another challenging problem is that of determining the shortest possible game that can be played on squares of order-13 or less. (Beyond 13, all 12 pieces have to be played, so that the problem becomes trivial.) In other words, what is the lowest cardinal number for a subset of the 12 pentominoes that can be played on an n-by-n board in such a way that no remaining piece will fit? Examples of the shortest games known on boards through order-13 are shown in Figure 78. In many cases more than one subset will yield a solution.

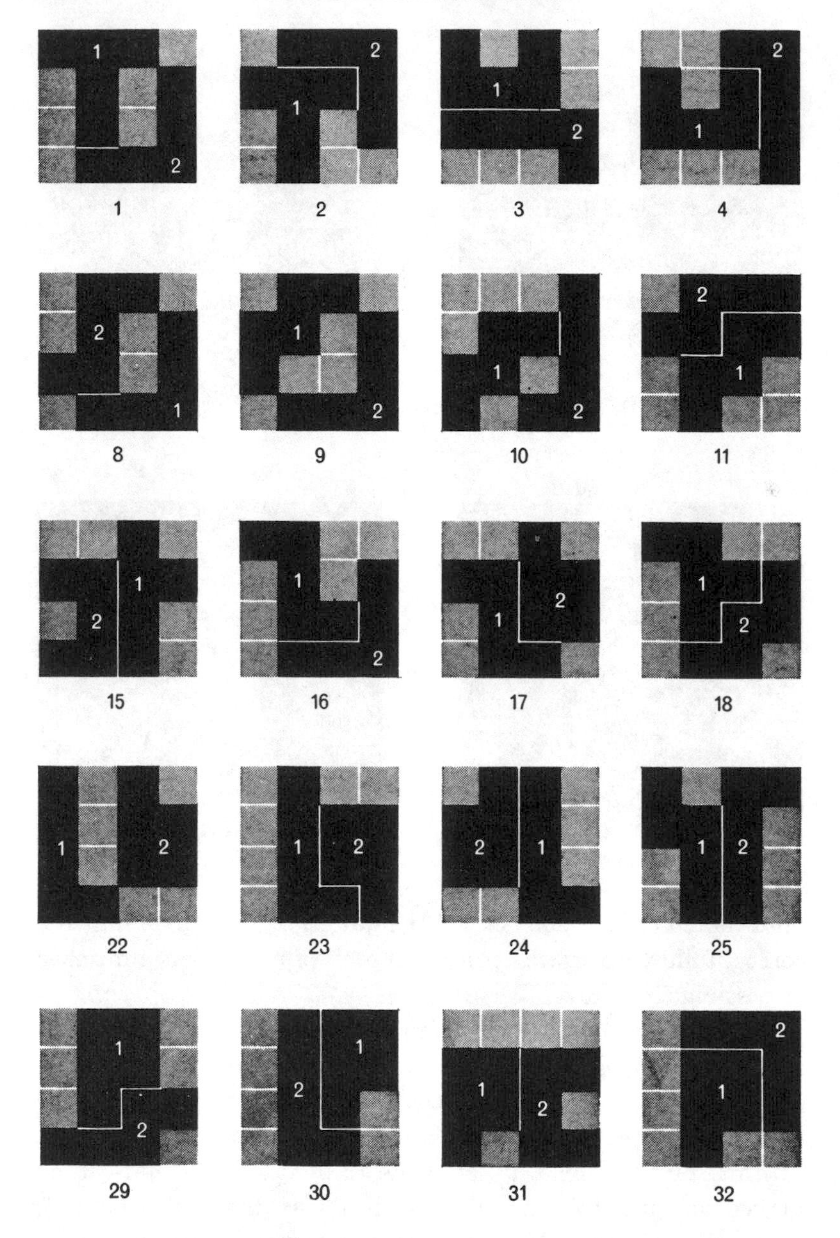
1
2
1
2
1
2
1
2
8
9
10
11
15
16
17
18
22
23
24
25
29
30
31
32

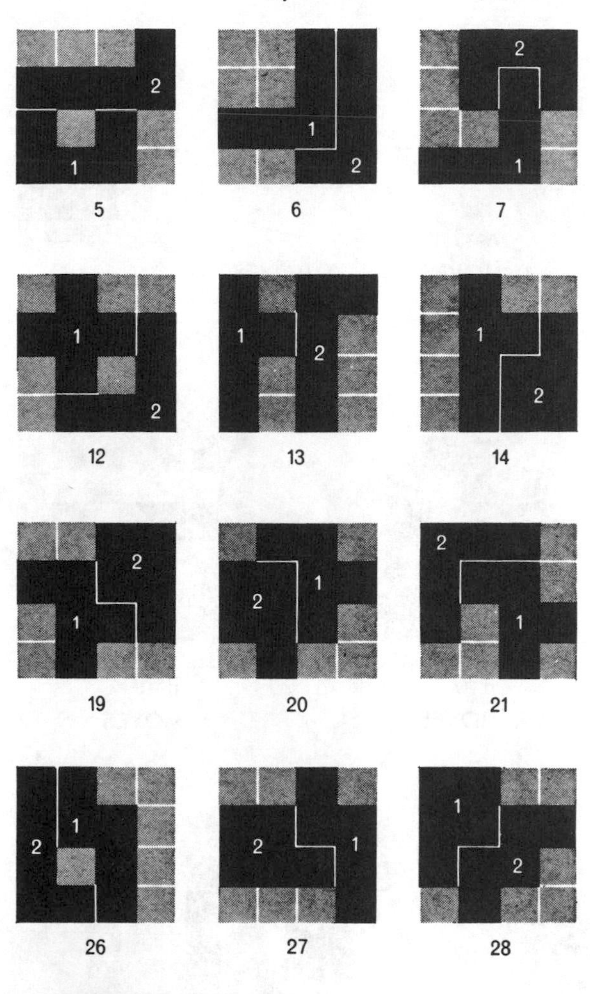

FIGURE 77
Proof of the second player's advantage on the four-by-four board

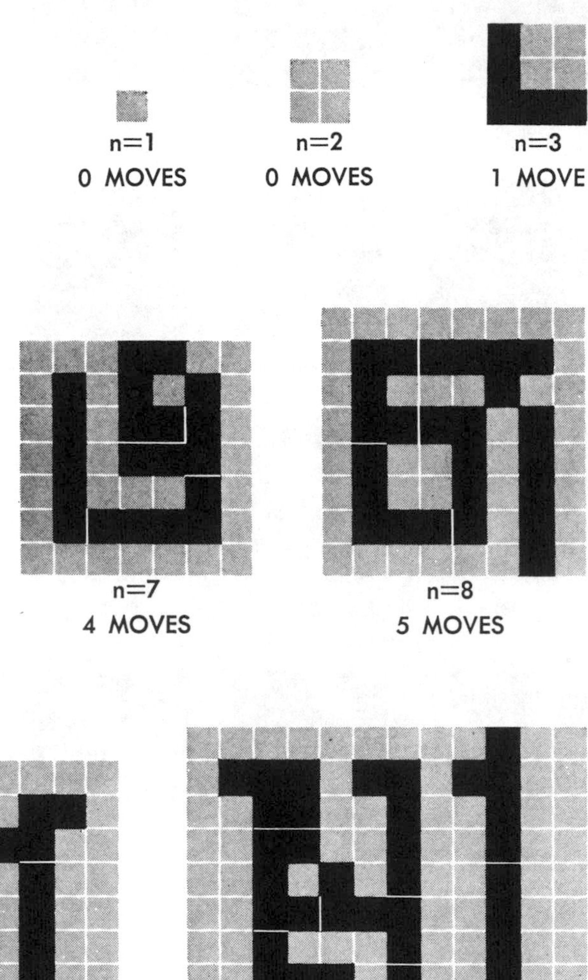

n=1
0 MOVES

n=2
0 MOVES

n=3
1 MOVE

n=7
4 MOVES

n=8
5 MOVES

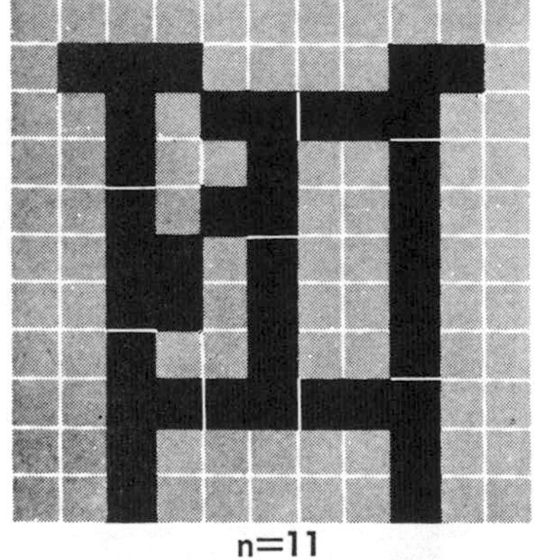

n=11
8 MOVES

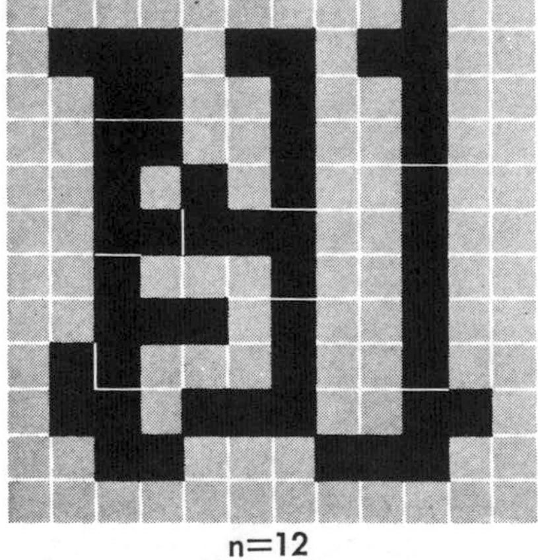

n=12
10 MOVES

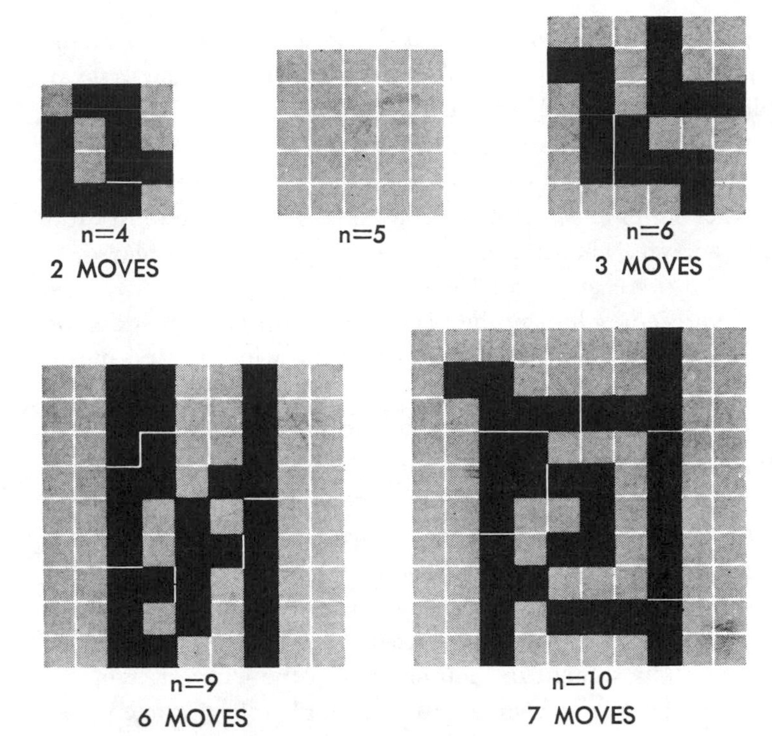

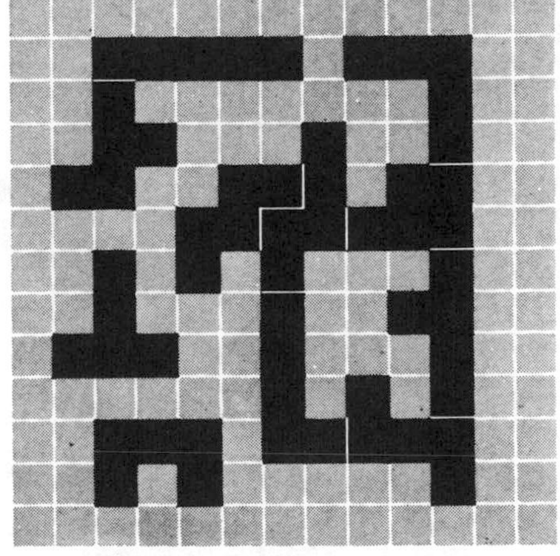

FIGURE 78
Shortest games known for boards of 1 to 13 squares per side

The shortest game on the 5-by-5 board has been left blank. Here are two easy problems. What is the shortest game that can be played on this field? What is the longest?

What about rectangular boards that are not square? In an exhaustive analysis of all such boards with areas of 36 square units or less, Golomb found that the 5-by-6 board was the most difficult to analyze. The first player can always win if he plays correctly, and there are several winning first moves. Readers who find the preceding four problems too easy may enjoy working on this much more difficult one: Find all the winning first moves on the 5-by-6 board.

A quite different type of polyomino problem—not discussed in Golomb's book and also far from fully explored—is that of determining if duplicates of a given polyomino will fit together to form a rectangle. (Asymmetric pieces may be turned over and placed either way.) If so, what is the smallest rectangle that can be so formed? If not, a proof of impossibility must be given. The problem was suggested by David Klarner when he was a graduate student in mathematics at the University of Alberta. The following year a group of high school students attending a summer institute in mathematics at the University of California at Berkeley studied the problem under the direction of their teacher, Robert Spira. They called it the "polyomino-rectification problem," using the term "rectifiable" for any polyomino that could be replicated to form a rectangle.

The monomino (one square) and domino (two squares) are obviously rectifiable since each is itself a rectangle. Both trominoes (three-square figures) are rectifiable: one is a rectangle, and two *L* trominoes form a two-by-three rectangle. Of the five tetrominoes (four-square shapes) the straight tetromino and the square tetromino are rectangles. Two *L* tetrominoes form a four-by-two, and the *T* tetromino replicates to fill the four-by-four square as shown in Figure 79*a*. The remaining tetromino is not rectifiable. The proof is trivial. If it is placed to fit the upper left corner of a rectangle, it is impossible to form a top edge that terminates at a second corner, as shown in Figures 79*b* and 79*c*.

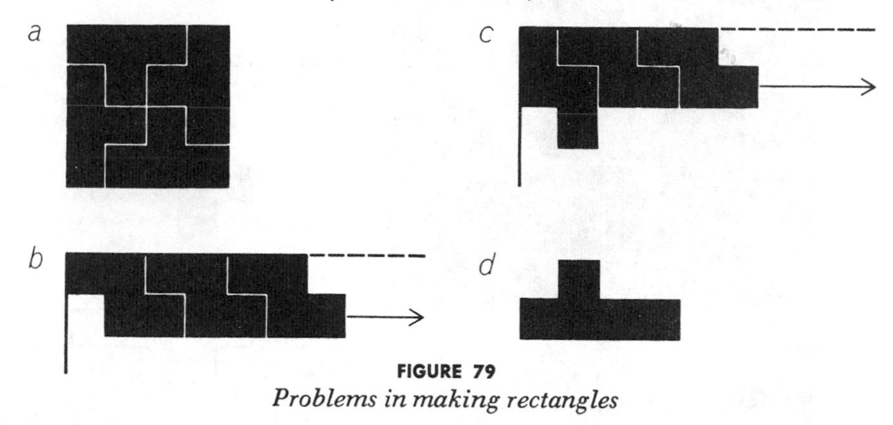

FIGURE 79
Problems in making rectangles

Similar impossibility proofs are easily found for most of the pentominoes. The reader may enjoy showing that the *T*, *U*, *V*, *W*, *X*, *Z*, *F*, and *N* pentominoes are not rectifiable. The *I*, *L*, and *P* are easily rectified. This leaves only the *Y*, the most difficult of the pentominoes to analyze. Is the *Y* pentomino, in Figure 79*d*, rectifiable? If so, what is the smallest rectangle it will fill? If not, prove it.

David Klarner has established that nine hexominoes are rectifiable. The only hexomino known to require more than four replications to form a rectangle is the piece shown, with its minimal pattern, at the top of Figure 80. Only one hexomino has not yet been proved to be either rectifiable or not. It is the hexomino shown at the top of Figure 81.

The only heptomino known to require more than four replicas to make a rectangle is the one shown at the bottom of Figure 80. Its smallest rectangle (discovered by James E. Stuart, of Endwell, New York) is known to be the order-14 square, requiring 28 pieces. If the reader will cut 28 replicas of this heptomino from cardboard or thin wood, he will find it a splendid puzzle to fit them into a square. (Pieces may be turned over and placed with either side up.) One heptomino, at the bottom of Figure 81, is not yet known to be rectifiable or not.

In a letter written in 1974, Klarner passed along a beautiful result which he obtained:

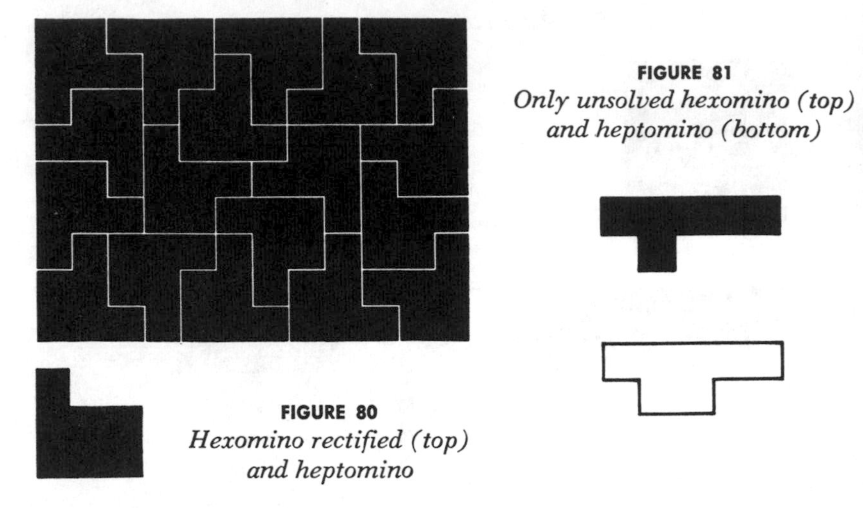

FIGURE 80
Hexomino rectified (top) and heptomino

FIGURE 81
Only unsolved hexomino (top) and heptomino (bottom)

Let R *denote the set of all sizes of rectangles which can be filled with copies of a particular* n-omino. *For example,* R *for the* Y *pentomino begins:* 5×10, 10×10, 10×14, 10×16, . . . *Then there exists a finite subset* S *of* R *such that every element in* R *can be cut into pieces each belonging to* S. *This means that there are only a finite number of "atomic" problems for a given* n-omino. *In other words, if we have packed enough sizes of rectangles, then all larger packable rectangles can be cut into rectangles of the smaller size. The set of atomic rectangles for the* Y *pentomino is not yet completely determined, although it is known up to a finite computation. In general, we are not sure there is a finite computation which determines the finite basis* S *of a given set* R. *Probably not.*

Klarner also reports that there is an algorithm which decides in a finite number of steps whether copies of a given finite set of polyominoes pack a $k \times n$ rectangle for some n where k is a given (fixed) natural number. There is, therefore, a procedure for answering: Does the set of polyominoes pack a $1 \times n$ rectangle? Does it pack a $2 \times n$ rectangle? Does it pack a $3 \times n$ rectangle? And so on. However, the question of whether the given set packs a rectangle is not decidable, though nobody seems to know if it is decidable whether replicas of a *single* polyomino will pack a rectangle.

ADDENDUM

GOLOMB'S PENTOMINO game was first described in my *Scientific American* column, November 1957 (reprinted in the first collection of columns). Since then, two unauthorized versions of the game have been marketed. The first, called Pan-Kāi (Phillips Publishers, Newton, Mass., 1960), used an order-10 board and a set of 12 pentominoes for each of two players. In 1967, Parker Brothers brought out Universe. The field was shaped like a fat cross, and four sets of pentominoes were included so that two, three, or four players could compete. As in Pan-Kāi, a rule prevents placing any piece so as to create an enclosed area with fewer than five cells. The box's cover displayed a scene from the motion picture *2001: A Space Odyssey* showing the game played on a computer aboard the spaceship. When the movie was later released, however, this episode had been replaced by a computer chess game.

The first authorized version of the game, with an instruction booklet prepared by Golomb, was marketed in 1973 by the Springbok Division of Hallmark Cards, just twenty years after Golomb introduced polyominoes to mathematicians in his memorable talk to the Harvard Mathematics Club.

ANSWERS

THE FIRST PROBLEM was to pick, from among the 33 different two-move pentomino games on the 4-by-4 board, the game that has only one winning reply by the second player. It is the game that was numbered 26 and is shown again in Figure 82*a*. The first play leaves a space on the right that can be filled only by the *L* pentomino. But if the *L* is placed in that space, the first player can win by playing on the left. If, on the other hand, the second player puts any piece except the *L* on the left, the first player can win by playing the *L* on the right. To win, therefore, the second player must place the *L* on the left as shown.

On the 5-by-5 board the first player has an obvious win by playing the *I* pentomino in the center, as shown at *b*. His oppo-

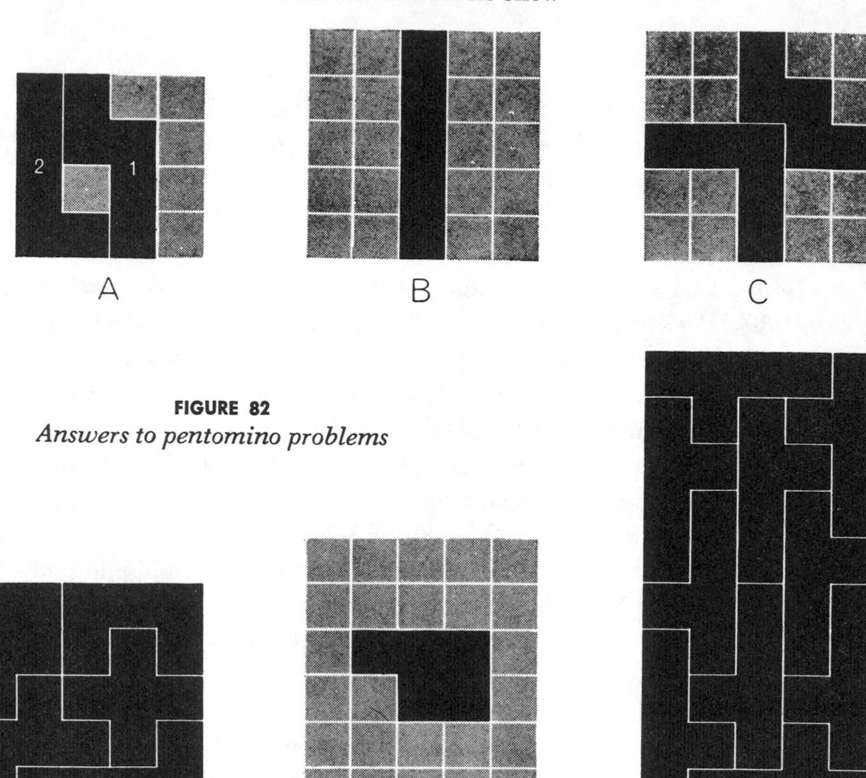

FIGURE 82
Answers to pentomino problems

nent must play on one side and the first player then wins by playing on the other side. The shortest possible game on this board has two moves (*c*) and the longest has five (*d*). The short-game pattern is unique, but there are many solutions to the long game.

The first player can win on the 5-by-6 board if his first move is the one shown at *e*. There is no simple proof, and space does not allow showing correct responses to all possible second moves. There are at least three other first-move wins.

The *Y* pentomino is rectifiable. The drawing at *f* shows the smallest rectangle that can be formed with replicas of this piece. The pattern illustrated is one of four possible solutions.

Is it possible to pack a rectangle of *odd* area with the *Y* pentomino? The answer is yes, and the smallest such rectangle is the order-15 square, found by Jenifer Haselgrove, a computer scientist at the University of Glasgow. It is a remarkably difficult problem, which I shall here leave unanswered.

James E. Stuart's rectification of the heptomino shown at the bottom of Figure 80 is given in Figure 83. It is not unique, because the four central heptominoes can be put together a different way, and each of the shaded pairs can be mirror-reversed. Note the pattern's pleasing fourfold symmetry.

Only four other heptominoes are known to be rectifiable. Their trivial minimum solutions are shown in Figure 84.

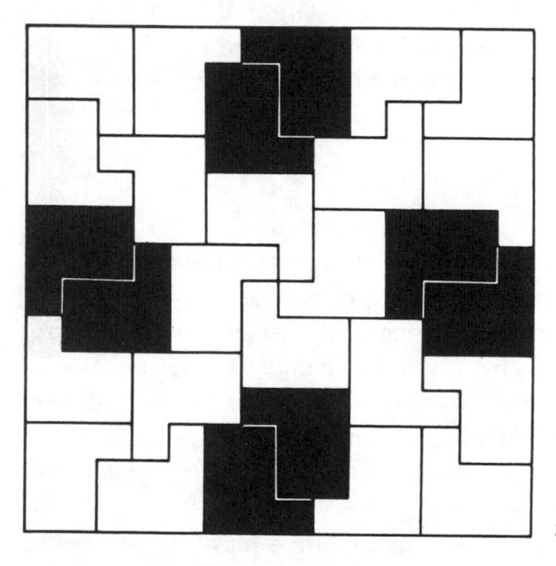

FIGURE 83
A square heptomino rectification

FIGURE 84
Four trivial heptomino rectifications

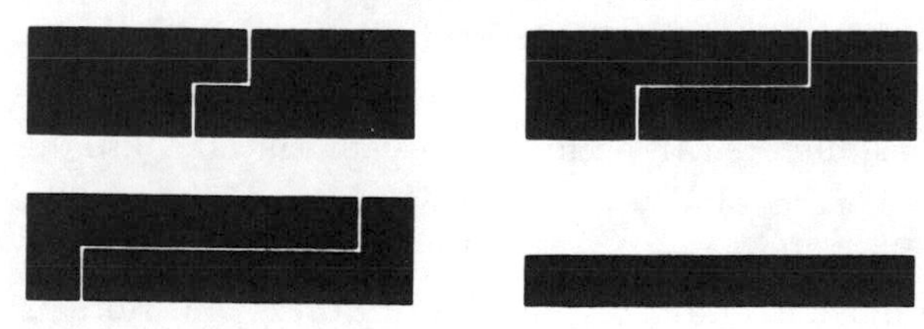

CHAPTER 14

Knights of the Square Table

He sat leaning on his cane and thinking that with a Knight's move of this lime tree standing on a sunlit slope one could take that telegraph pole over there . . .
—VLADIMIR NABOKOV, *in* The Defense

THE DEFENSE, a novel about a chess grand master, is not the only novel by Nabokov—himself a good chess player and composer of chess problems—in which characters see knight's moves in patterns around them. Humbert Humbert, the narrator of *Lolita*, observes a latticed window with one red pane and comments: "That raw wound among the unstained rectangles and its asymmetrical position—a knight's move from the top—always strangely disturbed me."

The knight is the only chess piece with a move that covers an asymmetrical pattern of squares; surely it is this lopsidedness that gives the move its disturbing strangeness. *Der Springer*, as the piece is called in German, springs two squares along a row or file and then, like Lewis Carroll's White Knight behind the mirror, topples one square either left or right. Another way of describing this asymmetrical gallop is to say that the knight moves one square orthogonally, like a rook, pivots 45 degrees to the left or right, and moves one square diagonally, like a bishop. This is how the move of the *ma* (horse) in Chinese and Korean chess must be explained because, unlike its Western counterpart, a *ma* cannot move if another piece occupies the diagonally adjacent square on which the pivot occurs. The *keima* (honorable horse) of Japanese chess moves like the Western knight, vaulting all pieces in its way, but it can only go forward across the board.

"The knight," said British puzzle expert Henry Ernest Dudeney, "is the irresponsible low comedian of the chessboard." No other chess piece has been the basis for so many unusual and amusing combinatorial problems. In this chapter we shall glance at a few of the classics along with some new discoveries by Solomon W. Golomb.

The oldest of knight puzzles, now the subject of an enormous literature, is the knight's tour. The problem is to find a single path of knight's moves (on boards of various sizes and shapes) that allows the knight to occupy each square once and only once. The tour is closed if the knight returns to its starting cell, open if the ends of the tour cannot be linked by a knight's move. If the board is checkerboard-colored, the colors of the cells will alternate along any tour. On a closed tour, therefore, there must be the same number of black cells as there are white. Since all odd-order square boards contain an odd number of cells, it follows that no closed tours are possible on such boards. Tours of both types are impossible on squares of sides 2 and 4 but exist on all higher squares of even order. The 3-by-4 is the smallest rectangle on which an open tour is possible, and the 5-by-6 and 3-by-10 are the smallest on which closed tours can be made. No tour of either type can be made if one side is less than 3, and no closed tour is possible if one side is 4.

The power of color patterns to provide short, elegant proofs of tour impossibilities is strikingly demonstrated by Golomb's method of showing that a closed tour is impossible on any rectangle of side 4. The 4-by-n board is labeled with four letters [*Figure 85*]. Observe that every A cell on a knight's path must be preceded and followed by a C cell. There are equal numbers of A and C cells, and all must lie on any closed tour. But the only way to catch all of them is by avoiding the B and D cells

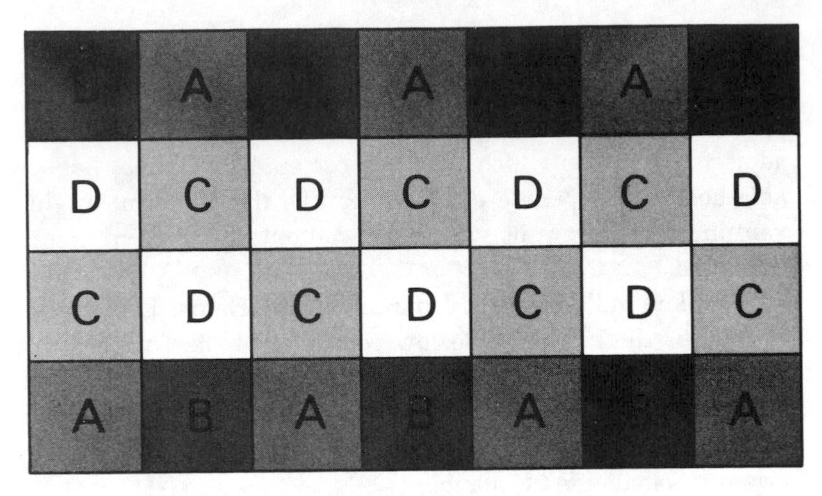

FIGURE 85
Shading for the 4-by-n board to prove the impossibility of a knight's closed tour

altogether, because once a leap is made from a *C* to a *D* cell there is no way to get back to an *A* cell without first landing on another *C* cell. If there is a closed tour, therefore, it will contain more *C* cells than *A* cells, and since this cannot be the case we conclude that such a tour is impossible. (For a similar proof, see Vol. 1, page 389, of the Ahrens work listed in the Bibliography.)

No one knows how many different knight's tours exist on the order-8 chessboard; varieties of one type of tour alone run into the millions. The search has usually been for tours that display unusual symmetry or that create a matrix (when cells along it are numbered consecutively) with remarkable arithmetical properties. For example, the closed tour shown in Figure 86, one of many constructed by Leonhard Euler in 1759, first covers the board's lower half, then its upper half, and all symmetrically opposite pairs of numbers (on a straight line through the center) have a difference of 32.

A closed tour with fourfold symmetry (the pattern is the same for all 90-degree rotations) is not possible on the order-8 board (or on any board with a side exactly divisible by 4). There are, however, five such tours on the order-6 board. The

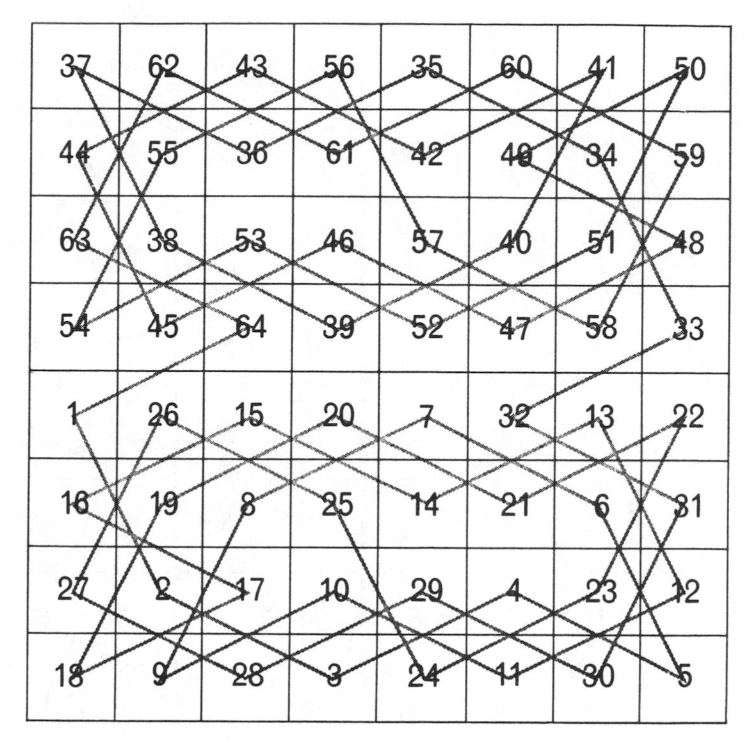

FIGURE 86

A closed tour by Euler. Symmetrically opposed pairs of numbers have difference of 32

interested reader will find them reproduced on page 263 of Maurice Kraitchik's *Mathematical Recreations* (Dover, 1953).

Figure 87 shows an open tour published by William Beverly in *The London, Edinburgh, and Dublin Philosophical Magazine and Journal of Science* for August 1848. (Whether this was William Roxby Beverly, a distinguished English landscape painter and stage designer of his time, I have been unable to determine.) Beverly's tour was the first "semimagic" tour to be constructed: the sum of each row and each file is 260. (The fact that the two main diagonals do not have the constant sum prevents the square from being "fully magic.") If Beverly's square is quartered as shown by the solid lines, each order-4 square is magic in rows and files, and if these four squares are in turn quartered, each order-2 square contains four numbers that add

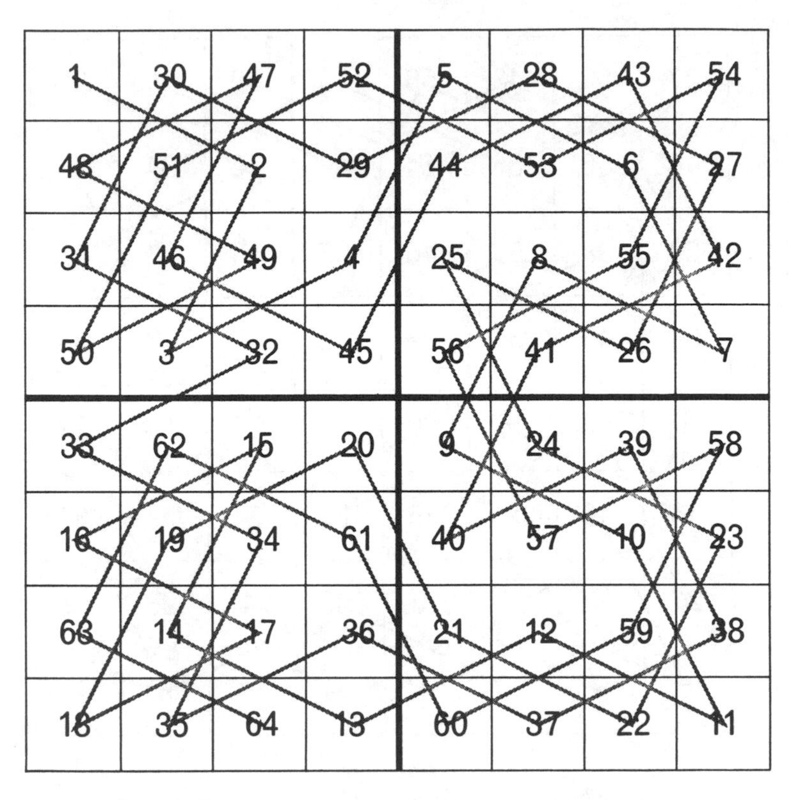

FIGURE 87

The first semimagic knight's tour: an open tour with each row and column adding to 260

to 130. Numbering the cells in reverse order along the tour produces the complement of the square: a semimagic square with all the properties of the former one.

Is there a fully magic knight's tour on the chessboard? That is the biggest unanswered question in knight's-tour theory. Scores of semimagic tours have been found, both open and closed, but none with even one main diagonal that has the required sum. It can be proved that fully magic tours are possible only on squares with sides that are multiples of 4. Since no tour is possible on the order-4, the chessboard is the smallest square for which the question is still open. Nor is a fully magic tour known for the order-12 square. Such tours have been constructed, however, for orders 16, 20, 24, 32, 40, 48, and 64. (A closed fully magic tour of order 16 is given on page 88 of Joseph S. Madachy's *Mathematics on Vacation*, 1966.)

What is the largest number of knights that can be placed on a chessboard so that no two attack each other? Intuitively one sees that the answer is 32, achieved by putting knights on all the black squares or on all the white. Proving it is a bit tricky. One way is to divide the board into 2-by-4 rectangles. A knight on any cell of such a rectangle can attack only one other knight, and so the rectangle cannot hold more than four nonattacking knights. Since there are eight such rectangles, no more than 32 nonattacking knights can go on the chessboard.

Golomb points out that a cleverer proof (contributed by Ralph Greenberg to *American Mathematical Monthly* for February 1964, page 210) rests on the existence of a knight's tour of the chessboard. As we have seen, along such a tour the colors of the cells alternate. Clearly we can place no more than 32 nonattacking knights on such a path. Equally obvious is the fact that they must go on alternating cells—that is, on all the white or all the black cells. Put another way, if we could place 33 nonattacking knights on the chessboard, any knight's tour would then have to include a hop from one cell to another of the same color, which is impossible. The mere existence of the tour not only proves that 32 is the maximum but also adds a surprise bonus: it proves the uniqueness of the two solutions. The proof generalizes to all even-order squares on which a tour is possible. On odd-order squares tours must of course begin and end on the same color. On such squares there is therefore only one solution: placing the knights on all cells that are the same color as the central square.

Turning from maxima to minima, let us ask: What is the smallest number of knights that can be placed on a square board so that all unoccupied cells are under attack by at least one knight? The following table gives the answers for boards of sides 3 to 10, and also gives the number of different solutions for each board (not counting rotations and reflections).

ORDER	PIECES	SOLUTIONS
3	4	2
4	4	3
5	5	8
6	8	22
7	10	3
8	12	1
9	14	1 (?)
10	16	2 (?)

Examples of solutions for orders 3 through 8 are shown in Figure 88. The unique chessboard solution has often been published. Patterns for the two next-higher boards, orders 9 and 10, are

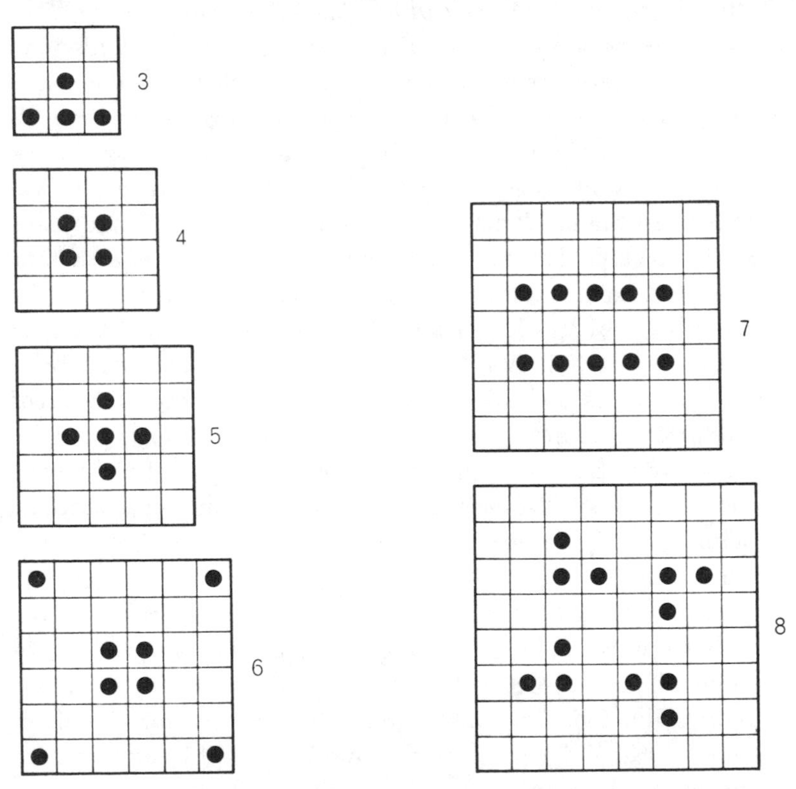

FIGURE 88
Solutions for boards of orders 3 through 8

not well known; the question marks in the table show that the order-9 solution is believed to be unique, and the order-10 is thought to have only two solutions. No knight, in any of the three patterns, is attacked by any other knight. Readers are invited to search for the solutions.

Note that in the pattern for order 7 all occupied cells are attacked, and in the order-8 pattern four occupied cells are attacked. If we add the proviso that *only* unoccupied cells be attacked, more knights are required for each of these boards. The best results known to me are 13 knights for order 7 and 14 for order 8. (I am indebted to Victor Meally for supplying order-8 solutions.)

Figure 89 left shows how 22 knights can be placed on the order-11 board so that all unoccupied cells are attacked, and also no knight is attacked. It was published in *L'Intermédiare des Mathématiciens*, Paris, Vol. 5, 1898, pages 230–31. This was believed to be a minimal solution, even when knights are permitted to attack other knights, until 1973 when Bernard Lemaire, of Paris, found the remarkable 21-knight solution shown in Figure 89 right. It was first published in the *Journal of Recreational Mathematics*, Vol. 6, Fall 1973, page 292.

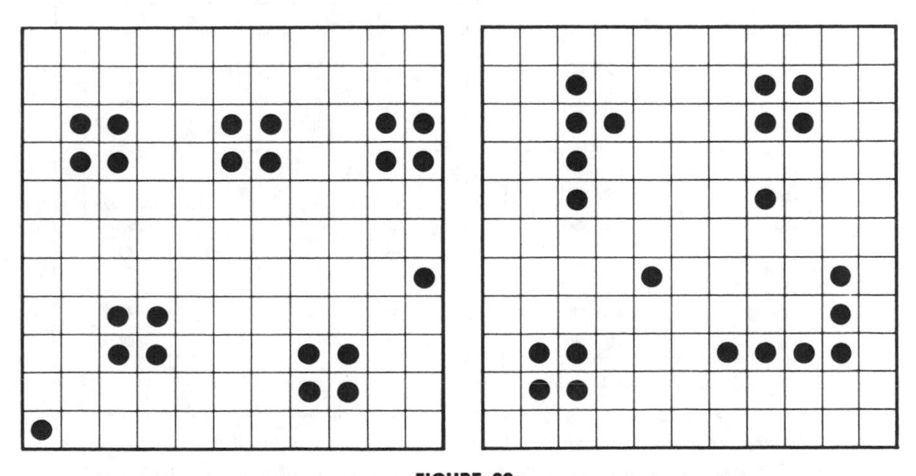

FIGURE 89

22 knights attack only *unoccupied cells (left), but 21 knights can attack all unoccupied cells (right), on the order-11 board*

All unoccupied cells of the order-12 board can be attacked with 24 knights as shown in Figure 90. This is also the best solution known if no knight is allowed to attack another knight. It is believed to be unique for both problems.

In Ahrens' work (Vol. 2, page 359), he gives the best-known solutions for attacking all unoccupied cells (knights may or may not attack one another) on boards of side 13, 14, and 15 as 28, 34, and 37 knights, respectively. In 1967 Harry O. Davis lowered the order-14 record to 32 knights. His bilaterally symmetric pattern (Figure 91) is here published for the first time.

If we ask that *all* cells, occupied or not, be attacked, the simplest approach is to draw two patterns on transparent paper, one for the minimum number of knights that attack all black cells and the other for the minimum number that attack all white cells. The two patterns can then be superposed in various ways to obtain final solutions. On the chessboard—as Dudeney explains in his solution to problem 319 in *Amusements in Mathematics*—there are only two patterns of seven knights (the minimum) that attack all cells of one color. By combining the two patterns in all possible ways to attack all 64 cells, one can

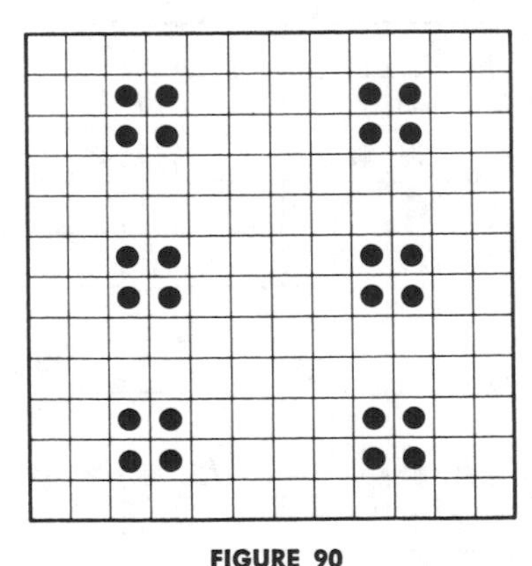

FIGURE 90
24 knights are minimal for attacking unoccupied cells on order-12 board

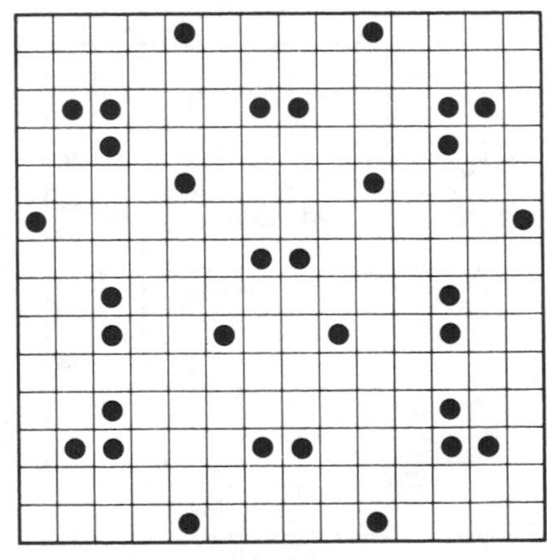

FIGURE 91
32 knights attack all unoccupied cells on order-14 board

obtain only three 14-knight patterns, not counting rotations and reflections as being different. I know of no work on this version of the problem on higher-order boards.

CHESKERS

ABOUT 1947 Golomb invented a hybrid game combining features of chess and checkers, which he naturally called "cheskers." Like checkers, it is played on the 32 black squares of the order-8 board. Since the knight cannot move on such a board without leaving the black squares, Golomb invented a modified knight that he recently christened the "cook." It moves three instead of two squares along a row or file, then one square at right angles. A centrally placed cook has eight moves [*see Figure 92*].

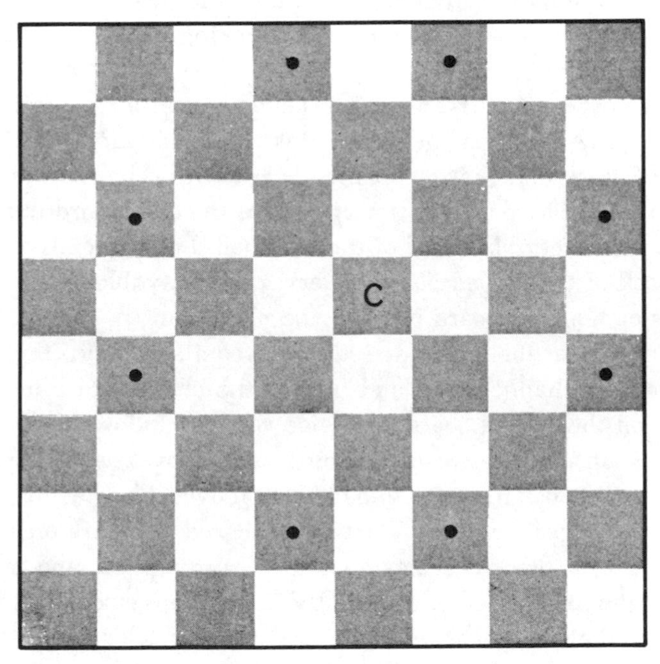

FIGURE 92
The eight moves of a centrally placed cook

The cook actually is the reinvention of a piece called the "camel" that was used in fourteenth-century Persian chess. The rules and board for this complex early version of chess are completely known because of a surviving Persian manuscript, the fullest translation of which is in Duncan Forbes's *History of Chess* (London, 1880). The game is known as Tamerlane's chess because Tamerlane the Great is supposed to have been fond of it. In addition to two camels on each side there are also two "asps" (corresponding to knights) and two powerful pieces called "giraffes" that move one cell diagonally and then continue forward orthogonally for any unblocked distance. Leonhard Euler worked on tours involving a piece that moved like the camel.

"The invention of the cook," Golomb writes, "immediately suggests two problems: Is there a cook's tour of the checkerboard? And how many cooks spoil the draughts? (That is, what is the maximum number of nonattacking cooks that can be placed on the board?)"

To answer the first question, Golomb uses a transformation of the chessboard suggested by his colleague Lloyd R. Welch [*see Figure 93*]. A jagged-edged board with cells twice the size of the chessboard cells is superposed on the chessboard in such a way that every black cell of the chessboard corresponds to a single cell of the jagged board. Every game playable on the black cells of the chessboard can now be played on the jagged board provided that the moves are suitably redefined. Since the transformation changes rows and files of the chessboard into diagonals on the jagged board, and vice versa, it follows that bishop moves on the chessboard become rook moves on the jagged board, and rook moves become bishop moves. Checkers is played on the jagged board by starting with red checkers on cells 1 through 12, black checkers on cells 21 through 32, and moving orthogonally instead of diagonally. (Has it ever occurred to the reader that, since checkers uses cells of one color only, two simultaneous but completely independent checker games can be played on the same checkerboard by four people seated around the board, each pair of opponents playing on a different color?)

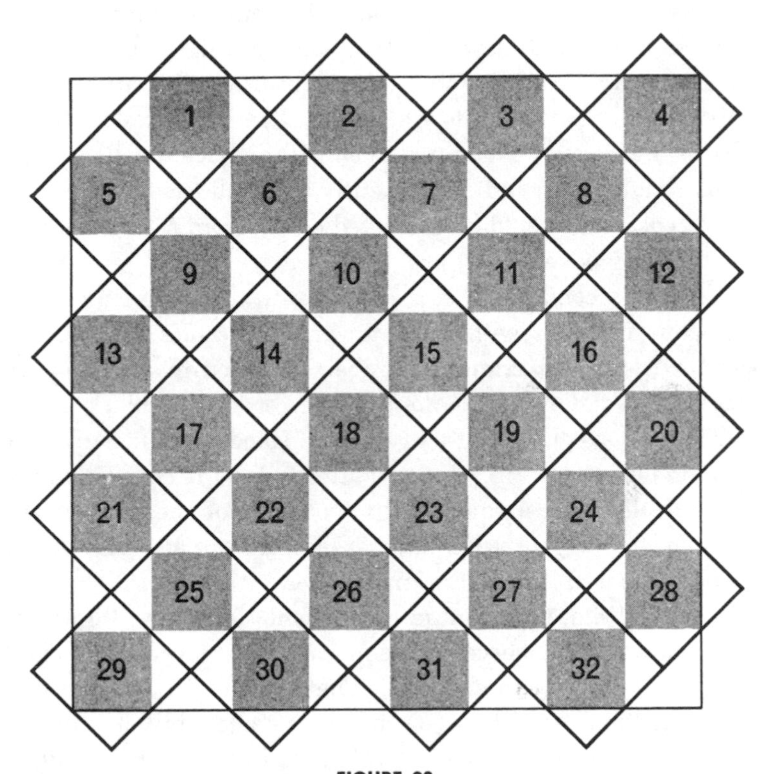

FIGURE 93
A transformation devised by Lloyd R. Welch

More surprisingly, Golomb points out, cook's moves on the chessboard turn into knight's moves on the jagged board! A cook's tour on the chessboard therefore corresponds to a knight's tour on the jagged board. A sample closed knight's tour on the jagged board is 1–14–2–5–10–23–17–29–26–32–20–8–19–22–9–21–18–30–27–15–3–6–11–24–12–7–4–16–28–31–25–13. Those numbers trace a cook's tour on the black cells of the chessboard. (For two other closed cook's tours on the standard chessboard, see Maurice Kraitchik, *Mathematical Recreations*, page 265.)

Since every cook's move on the chessboard joins two cells that are separated by two knight's moves, it occurred to Golomb that there might be a knight's tour of the chessboard of such a nature that every alternate cell along it would provide a cook's tour. He soon found, however, that when a knight enters a corner cell of the board, it jumps there from a cell that is diagonally adjacent to the cell to which it will be forced to leap when it leaves the corner. Those two diagonally adjacent cells are not a cook's move apart and consequently, Golomb writes, "the hope that a cook's tour could be extracted easily from a knight's tour is hopelessly cooked."

Golomb's second question is answered in the same way as the analogous problem with knights. Since a cook's tour of the board exists, the maximum number of cooks must occupy 16 alternate cells along such a tour. If the reader will mark the 16 even cells along the given tour (or the 16 odd cells), he will be marking one of the two solution patterns. On the chessboard the marked cells form a square lattice on half of the cells of one color. On the jagged board the marked cells are all those of one color if the board is checkerboard-colored.

For those who may want to try Golomb's cheskers, Figure 94 shows how the 12 pieces of each side are placed. The eight "men" (*M*) move as checkers. The two kings (*K*) move as checker kings. The bishop (*B*) moves as a chess bishop, and the cook (*C*) moves as previously explained. Like the chess knight, the cook is not obstructed by intervening pieces. The men and kings capture as in checkers, by leaping over the victim. The bishop and cook capture as in chess, by moving onto the square of the victim. If a checker capture exists, it is compulsory to make it, unless a chess capture is also available, in which case the chess capture may be made if one wishes. Chess captures are optional. A man that reaches the last row must be promoted, but the owner of the piece may choose to make it either a king, a bishop, or a cook.

Players alternate in moving. The object of the game is to capture all the opponent's kings. The first player with no kings is the loser. It is therefore an important strategic decision, Golomb writes, whether to promote a man to a king (for better defense) or to a bishop or cook (for offense). As in checkers, a blocked position is a loss for the player unable to move.

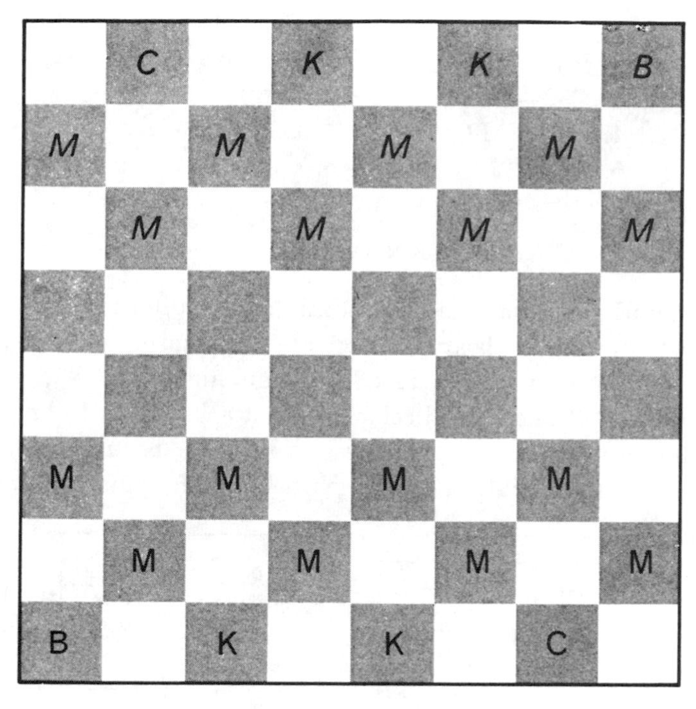

FIGURE 94
Starting position for Solomon W. Golomb's game of "cheskers"

ADDENDUM

RUFUS P. ISAACS pointed out that the jagged board shown in Figure 93 solves a puzzle he recalls seeing in the *New York World* when he was a boy. "A Scottish checker player," Isaacs wrote, "became irritated at the wastefulness of his board. He cut away half the squares. The remainder was in one simply connected piece on which it was still possible to play a legal game of checkers, using the original squares and making no additional markings. How did he do it?"

Several readers attempted to devise a cheskers' "fool's mate"; that is, the shortest possible legal game. The shortest came from Wilfred H. Shepherd, Manchester, England. Black squares are numbered as shown in Figure 93. *K*, *C*, *B*, *M* stand, respectively, for king, cook, bishop, and man. The game is:

1. M22—17	1. C1—13
2. M23—19	2. C13—18
3. C32—20	3. C18 × K30
4. C20 × M8	4. C30—18
5. C8 × K2	5. C18 × K31 (wins)

ANSWERS

FIGURE 95 shows how to place a minimum number of knights, 14, on the order-9 board so that all unoccupied cells are attacked. The solution is believed to be unique. Figure 95, right, shows how all unoccupied cells can be attacked with 16 knights on the order-10 board. This was thought to be unique until eleven readers found the second solution shown in Figure 96.

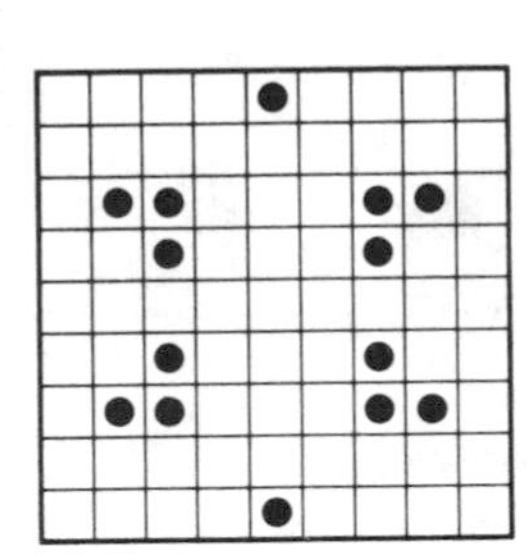
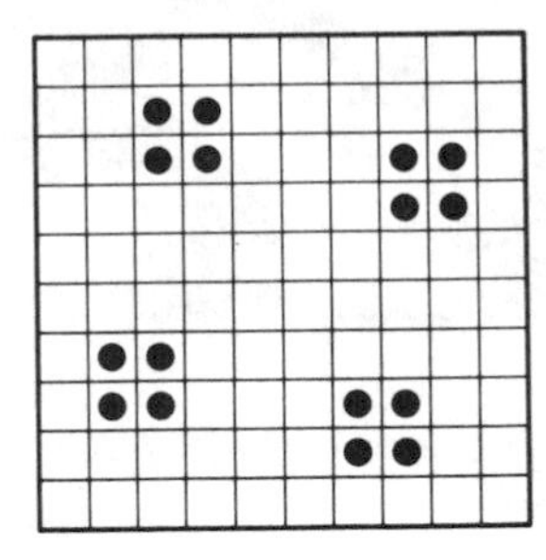

FIGURE 95
Solutions for boards of sides 9 and 10

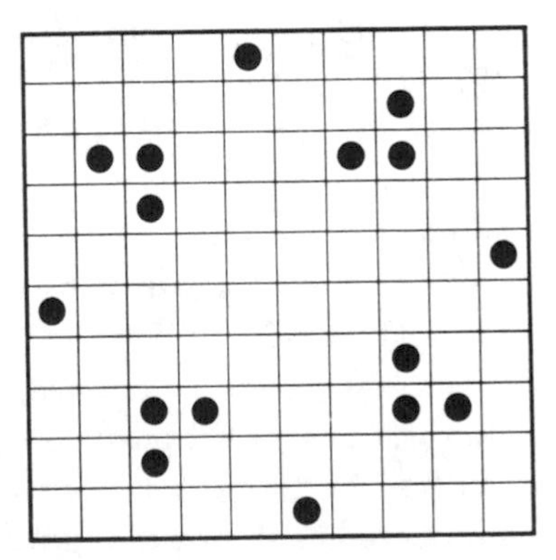

FIGURE 96
Newly discovered second solution for order-10 board

CHAPTER 15

The Dragon Curve and Other Problems

1. INTERRUPTED BRIDGE GAME

A TELEPHONE CALL interrupts a man after he has dealt about half of the cards in a bridge game. When he returns to the table, no one can remember where he had dealt the last card. Without learning the number of cards in any of the four partly dealt hands, or the number of cards yet to be dealt, how can he continue to deal accurately, everyone getting exactly the same cards he would have had if the deal had not been interrupted?

2. NORA L. ARON

A COLLEGE GIRL has the unusual palindromic name Nora Lil Aron. Her boyfriend, a mathematics major, bored one morning by a dull lecture, amuses himself by trying to compose a good number cryptogram. He writes his girl's name in the form of a simple multiplication problem:

$$\begin{array}{r} \text{NORA} \\ \text{L} \\ \hline \text{ARON} \end{array}$$

Is it possible to substitute one of the ten digits for each letter and have a correct product? He is amazed to discover that it is, and also that there is a unique solution. The reader should have little trouble working it out. It is assumed that neither four-digit number begins with zero.

3. POLYOMINO FOUR-COLOR PROBLEM

POLYOMINOES are shapes formed by joining unit squares. A single square is a monomino, two squares are a domino, three can be combined to make two types of trominoes, four make five different tetrominoes, and so on. I recently asked myself: What is the lowest order of polyomino four replicas of which can be placed so that every pair shares a common border segment? I believe, but cannot prove, that the octomino is the answer. Five solutions (there are more) were found by John W. Harris of Santa Barbara, California [*see Figure 97*]. If each piece is regarded as a region on a map, each pattern clearly requires four colors to prevent two bordering regions from having the same color.

Let us now remove the restriction to four replicas and ask: What is the lowest order of polyomino any number of replicas of which will form a pattern that requires four colors? It is not necessary for any set of four to be mutually contiguous. It is only necessary that the replicas be placed so that, if each is given a color, four colors will be required to prevent two pieces of the same color from sharing a common border segment. Regions formed between the replicas are not considered part of the "map." They remain uncolored. The answer is a polyomino that is of much lower order than eight.

4. HOW MANY SPOTS?

THIS DOUBLE PROBLEM was given by D. Mollison of Trinity College, Cambridge, in a 1966 problems contest for members of the Archimedeans, a Cambridge student mathematics society. The

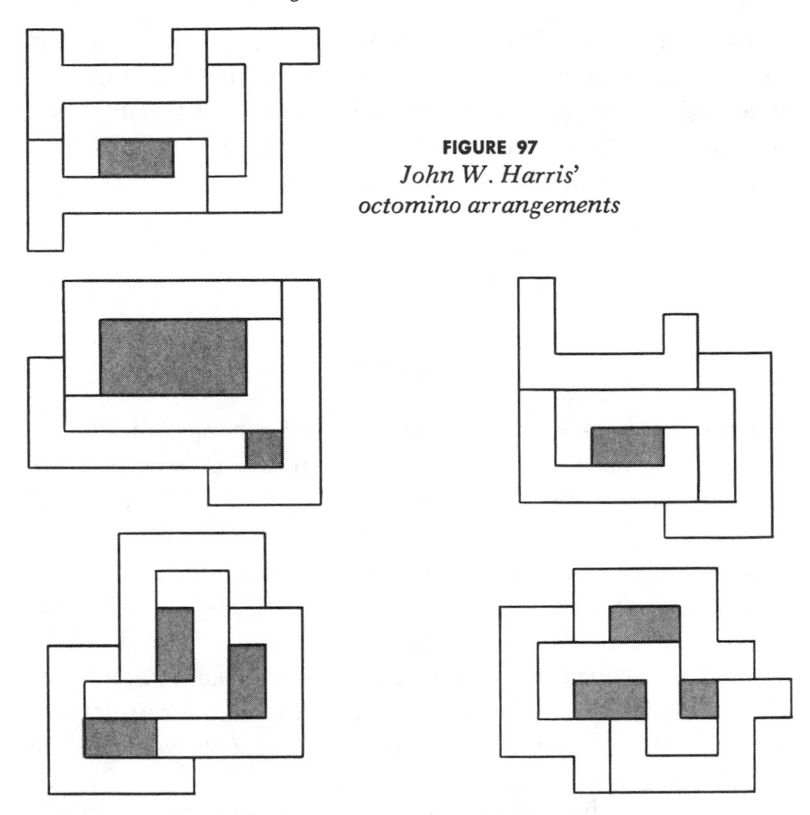

FIGURE 97
John W. Harris' octomino arrangements

first question: What is the maximum number of points that can be placed on or within the figure shown in Figure 98, provided that no two points are separated by a distance that is less than the square root of 2?

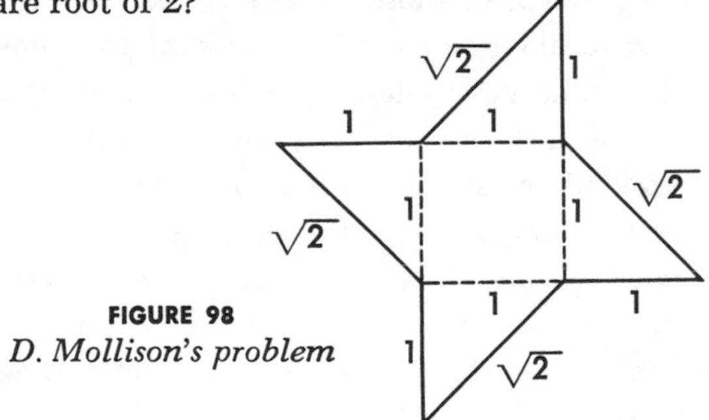

FIGURE 98
D. Mollison's problem

The second question: In how many different patterns can this maximum number of points be placed, not counting rotations and reflections of patterns as being different? The dotted lines were added to the figure to show that it is formed by a unit square surrounded by four half-squares.

5. THE THREE COINS

WHILE your back is turned a friend places a penny, nickel, and dime on the table. He arranges them in any pattern of heads and tails provided that the three coins are not all heads or all tails.

Your object is to give instructions, without seeing the coins, that will cause all three to be the same (all heads or all tails). For example, you may ask your friend to reverse the dime. He must then tell you whether you have succeeded in getting all the coins alike. If you have not, you again name a coin for him to turn. This procedure continues until he tells you that the three coins are the same.

Your probability of success on the first move is 1/3. If you adopt the best strategy, what is your probability of success in two moves or fewer? What is the smallest number of moves that guarantees success on or before the final move?

The reader should find those questions easy to answer, but now we complicate the game a bit. The situation is the same as before, only this time your intent is to make all the coins show heads. Any initial pattern except all heads is permitted. As before, you are told after each move whether or not you have succeeded. Assuming that you use the best strategy, what is the smallest number of moves that guarantees success? What is your probability of success in two moves or fewer, in three moves or fewer, and so on up to the final move at which the probability reaches 1 (certainty)?

6. THE 25 KNIGHTS

EVERY SQUARE of a 5-by-5 chessboard is occupied by a knight. Is it possible for all 25 knights to move simultaneously in such a way that at the finish all cells are occupied as before? Each move must be a standard knight's move: two squares in one direction and one square at right angles.

7. THE DRAGON CURVE

A WEIRD cover design decorated a booklet that William G. Harter, then a candidate for a doctorate in physics at the University of California at Irvine, prepared for a National Aeronautics and Space Administration seminar on group theory that he had taught the previous summer at NASA's Lewis Research Center in Cleveland [*see Figure 99*]. The "dragon curve," as he calls it, was discovered by a NASA colleague, physicist John E. Heighway, and later analyzed by Harter, Heighway, and Bruce A. Banks, another NASA physicist. The curve is not connected with group theory, but it was used by Harter to symbolize what he calls "the proliferation of cryptic structure that one finds in this discipline." It is drawn here as a fantastic path along the lattice lines of graph paper, with each right-angle turn rounded off to make it clear that the path never crosses itself. You will see that the curve vaguely resembles a sea dragon paddling to the left with clawed feet, his curved snout and coiled tail just above an imaginary waterline.

The reader is asked to find a simple method of generating the dragon curve. In the answer I shall explain three: one based on a sequence of binary digits, one on a way of folding paper, and one on a geometric construction. It was the second procedure that led to the discovery of the curve. I shall also explain the significance of the 12 spots, which indicate that this is a dragon curve of order 12. They happen to lie on a logarithmic spiral, although this was not noticed until later and it plays no role in the construction of the curve.

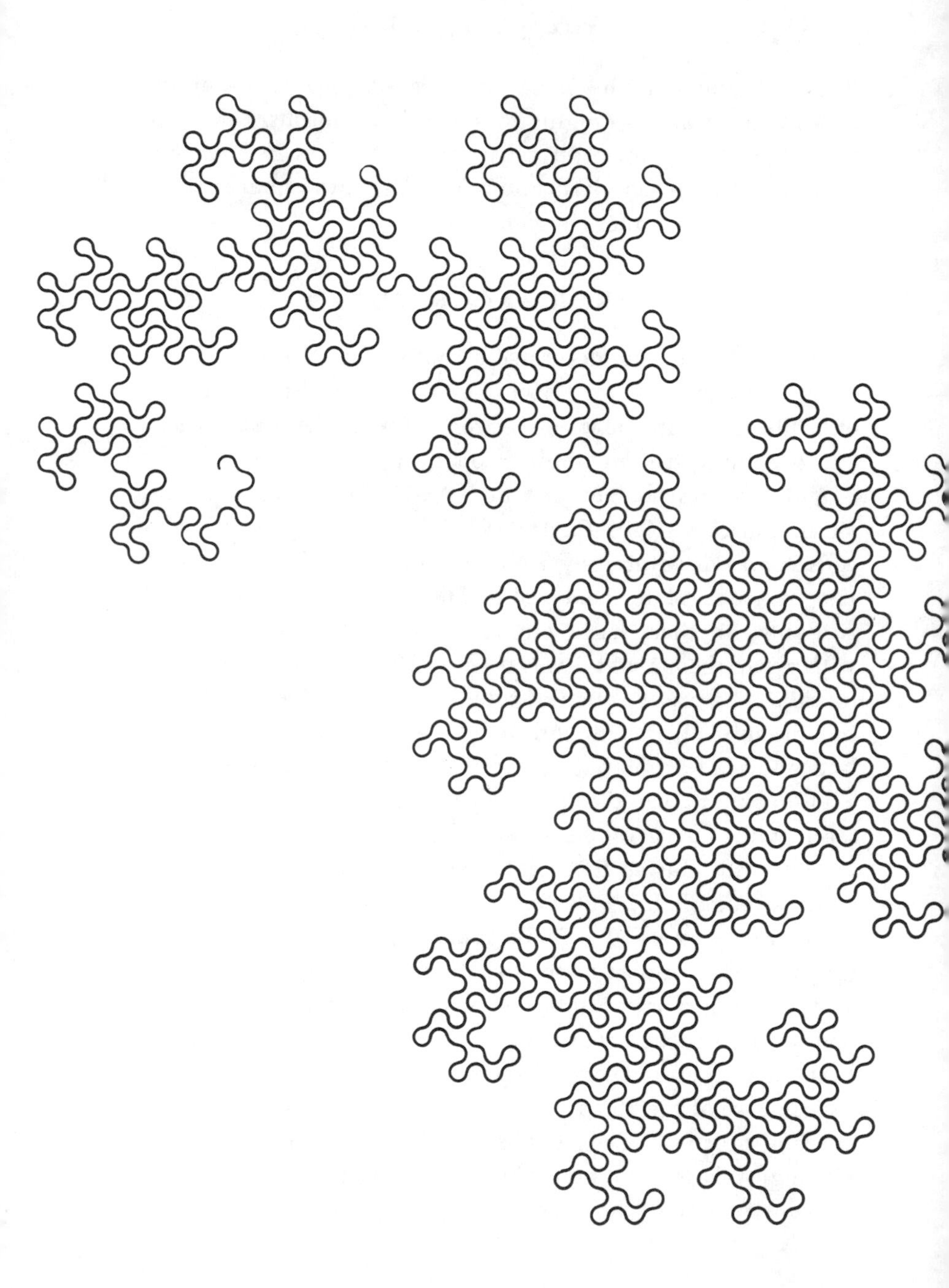

FIGURE 99
A "dragon curve" of the 12th order

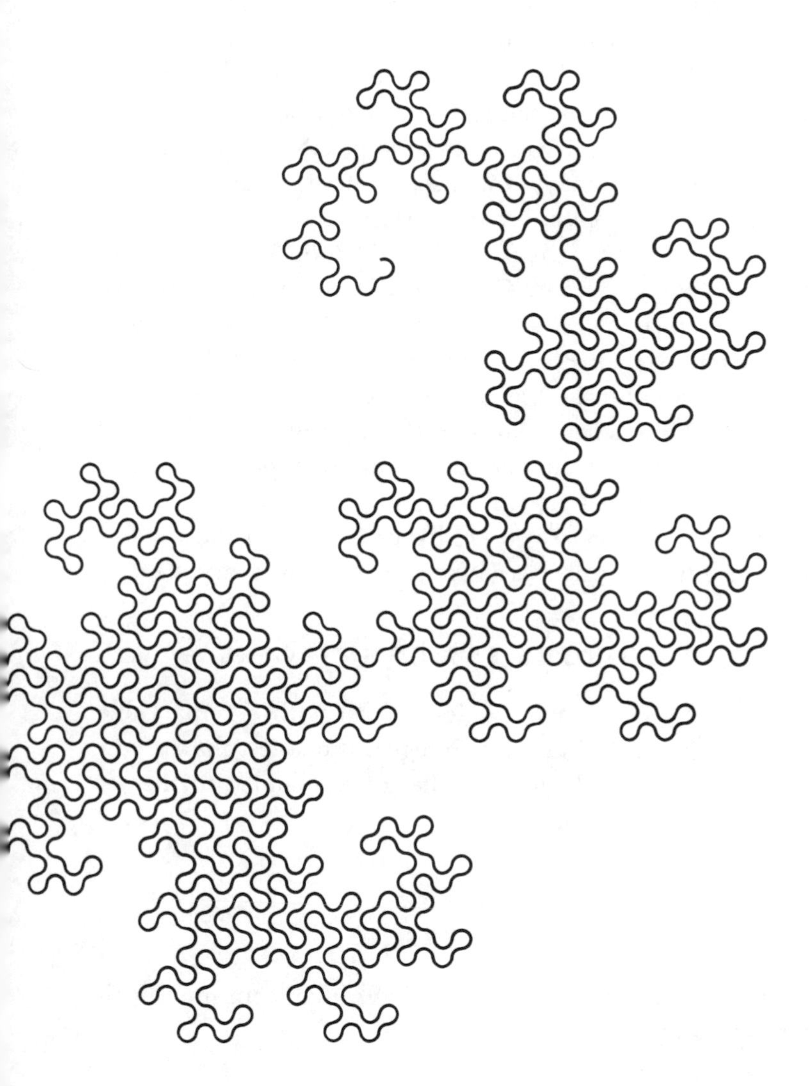

8. THE TEN SOLDIERS

TEN SOLDIERS, no two of them the same height, stand in a line. There are 10!, or 3,628,800, different ways the men can arrange themselves, but in every arrangement at least four soldiers will form a series of ascending or descending heights. If all but those four leave the line, the four will stand like a row of panpipes.

You can convince yourself of this by experimenting with ten playing cards bearing values from ace to 10. The values represent the height order of the soldiers. No matter how you arrange the ten cards in a row, it will always be possible to pick out at least four cards (there may, of course, be more) in ascending or descending order. Suppose, for instance, you arrange the cards in the following order: 5, 7, 9, 2, 1, 4, 10, 3, 8, 6. The set 5, 7, 9, 10 is in ascending order. Can you eliminate such a set by moving, say, the 10 to between the 7 and the 9? No, because you then create the set 10, 9, 8, 6, which is in *descending* order.

Let p (for panpipe) be the number of the largest set of soldiers that will always be found in order in a row of n soldiers of n different heights, no matter how they arrange themselves. The problem—and it is not easy—is to prove that if n equals 10, p equals 4. In doing so you are likely to discover the general rule by which the p number is easily computed for every n.

9. A CURIOUS SET OF INTEGERS

THE INTEGERS 1, 3, 8, and 120 form a set with a remarkable property: the product of any two integers is one less than a perfect square. Find a fifth number that can be added to the set without destroying this property.

ANSWERS

1. HE DEALS the bottom card to himself, then continues dealing from the bottom counterclockwise.

2. NORA × L = ARON has the unique solution 2178 × 4 = 8712. Had Nora's middle initial been A, the unique solution would have been 1089 × 9 = 9801. The numbers 2178 and 1089 are the only two smaller than 10,000 with multiples that are reversals of themselves (excluding trivial cases of palindromic numbers such as 3443 multiplied by 1). Any number of 9's can be inserted in the middle of each number to obtain larger (but dull) numbers with the same property; for instance, 21999978 × 4 = 87999912.

For a report on such numbers, in all number systems, see "Integers That Are Multiplied When Their Digits Are Reversed," by Alan Sutcliffe, in *Mathematics Magazine*, Vol. 39, No. 5, November 1966, pages 282–87.

Larger numbers can also be fabricated by repeating each four-digit number: thus, 217821782178 × 4 = 871287128712, and 108910891089 × 9 = 980198019801. Of course numbers such as 21999978 may also be repeated to produce reversible numbers. Leonard F. Klosinski and Dennis C. Smolarski, in their paper "On the Reversing of Digits," *Mathematics Magazine*, Vol. 42, September 1969, pages 208–10, show that 4 and 9 are the only numbers which can serve as multipliers for reversing non-palindromic numbers. This can be put another way. If an integer is a factor of its reversal, the larger of the two numbers divided by the smaller must equal 4 or 9.

The fact that 8712 and 9801 are the only four-digit numbers that are integral multiples of their reversals is cited by G. H. Hardy, in his famous *Mathematician's Apology*, as an example of nonserious mathematics. For those who are fascinated by such oddities, I pass along the following chart, sent by Bernard Gaiennie, which points up the curious relationship between the two numbers:

1089	6534
2178	7623
3267	8712
4356	9801
5445	

The nine numbers are, of course, the first nine multiples of 1089. Note the consecutive order of the digits when you go up and down the columns. If the first five numbers are multiplied respectively by 9, 4, $2\frac{1}{3}$, $1\frac{1}{2}$, and 1, the products give the last five numbers in reverse order. Now 1, 4, and 9 are the first three square numbers, but those other two multipliers, $2\frac{1}{3}$ and $1\frac{1}{2}$, seem to come out of left field!

3. Figure 100 shows how as few as six dominoes can be placed so that if each is given a color, four colors are necessary to prevent two dominoes of the same color from touching along a border.

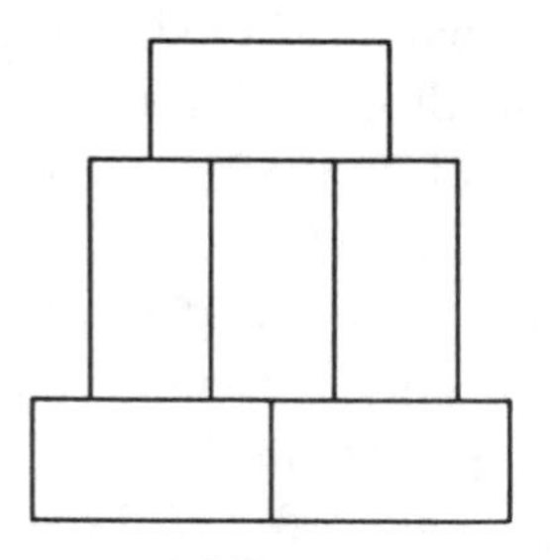

FIGURE 100
Six dominoes, four colors

The above is the solution I provided. To my astonishment, two readers (Bent Schmidt-Nielsen and E. S. Ainley) found a way of doing it with as few as eleven monominoes (unit squares). Their solution is shown in Figure 101. Informal proofs that eleven is minimal were supplied by R. Vincent Kron and W. H. Grindley.

If we ask for the maximum number of monominoes that can be placed so each pair shares a common edge segment, the answer obviously is three. The question is not so easily answered for cubes; that is, what is the maximum number of cubes that can be placed so each pair shares a common surface? A common surface need not be an entire face, but it must be a surface, not

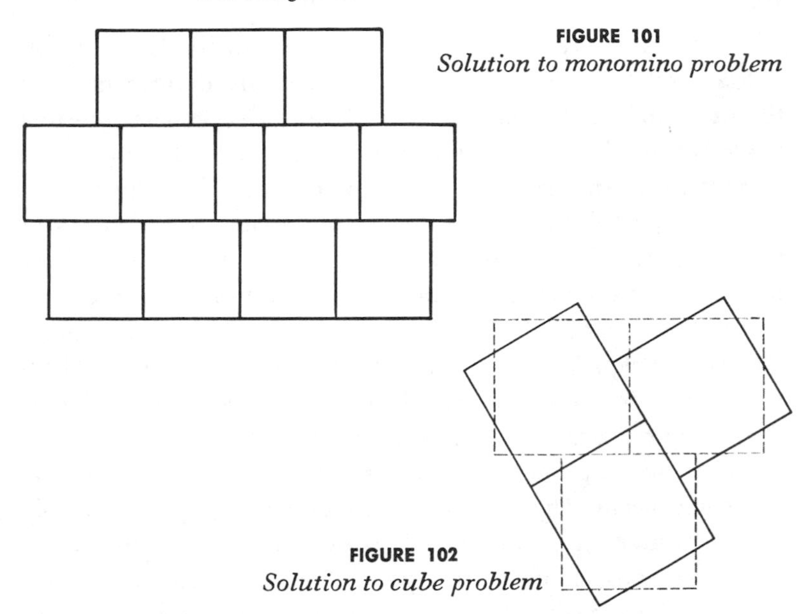

FIGURE 101
Solution to monomino problem

FIGURE 102
Solution to cube problem

a line or point. The answer is six. See Figure 102 for the pretty solution. The three cubes shown by the solid lines rest on three shown by dotted lines.

4. Five spots can be placed on the figure as shown in Figure 103 so that each pair is separated by a distance equal to the square root of 2 or more. There is enough leeway to allow each dot to be shifted slightly and therefore the number of different patterns is infinite. Did the reader fall into the carefully planned trap of thinking each spot had to fall on a vertex?

The problem appeared in *Eureka,* the journal of the Archimedeans, October 1966, page 19.

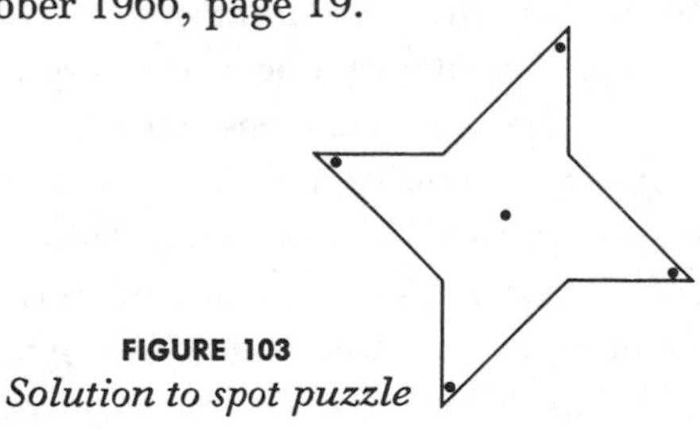

FIGURE 103
Solution to spot puzzle

5. The best way to make the three coins all heads or all tails is to direct that any coin be turned, then any other coin, then the first coin mentioned. The probability of success on the first move is 1/3. If you fail, the probability is 1/2 that your second move will do the trick. It might be supposed that the sum of those two probabilities is the chance of success in two moves or fewer, but this is incorrect. One must examine the effect of the first two moves on each of the six equally possible initial patterns, *HHT*, *HTH*, *HTT*, *THH*, *THT*, *TTH*. The symmetry allows one to pick any two coins for the first two moves. There is success in four cases, so that the chance of success on or before the second move is four out of six, or 2/3.

Seven moves guarantees success if the intent is to make all the coins heads. Of the eight possible starting patterns, only *HHH* is ruled out. You must therefore run through seven pattern variations to make sure you hit *HHH* somewhere along the line. An easily remembered strategy, suggested by Samuel Schwartz, is to label the coins 1, 2, 3 and take them in the order 1, 2, 3, 2, 1, 2, 3. The probability of success on the first move is 1/7, on or before the second move it is 2/7, and so on up to 7/7, or 1, on or before the seventh move.

If the number of coins is n, the required number of moves clearly is $2^n - 1$. The sequence of moves corresponds to the sequence of numbers in a binary Gray code. (On Gray codes, see my *Scientific American* column for August 1972.) Rufus Isaacs and Anthony C. Riddle each analyzed the situation as a competitive game. The player who hides the coins tries to maximize the number of moves, and the player who searches for the pattern tries to minimize the number of moves. The hider's best strategy is to pick randomly one of the seven possible states of the three coins. The searcher's best strategy is to label the corners of a cube with binary numbers 1 through 8, as explained in the August 1972 column, then draw a Hamiltonian path on the cube's edges. The sequence of moves corresponds to the sequence of binary numbers obtained by starting at the corner corresponding to 111 or 000, choosing with equal probability one of the two directions along the path, then traversing the path. If both players play optimally, the expected number of moves is four.

The general problem, given in terms of n switches that turn on a light only when all switches are closed, appeared in *American Mathematical Monthly*, December 1938, page 695, problem E319. For the general theory of search games of this type, see Rufus Isaacs, *Differential Games* (Wiley, 1965), pages 345 f.

6. The 25 knights cannot simultaneously jump to different squares. This is easily proved by a parity check. A knight's move carries the piece to a square of a different color from that of the square where it started. A 5-by-5 chessboard has 13 squares of one color, 12 of another. Thirteen knights obviously cannot leap to 12 squares without two of them landing on the same square. The proof applies to all boards with an odd number of squares.

If rooks are substituted for knights, but limited to a move of one square, the same impossibility proof obviously applies. It would also apply, of course, to any mixture of knights and such rooks.

7. Each dragon curve can be described by a sequence of binary digits, with 1's standing for left turns and 0's for right turns as the curve is traced on graph paper from tail to snout. The formula for each order is obtained from the formula for the next lowest order by the following recursive technique: add 1, then copy all the digits preceding that 1 but change the center digit of the set. The order-1 dragon has the formula 1. In this case, after adding a 1 there is only one digit on the left, and since it is also the "center" digit we change it to 0 to obtain 110 as the order-2 formula. To get the order-3 formula add 1, followed by 110 with the center digit changed: 1101100. Higher-order formulas are obtained in the same way. It is easy to see that each dragon consists of two replicas of dragons of the next

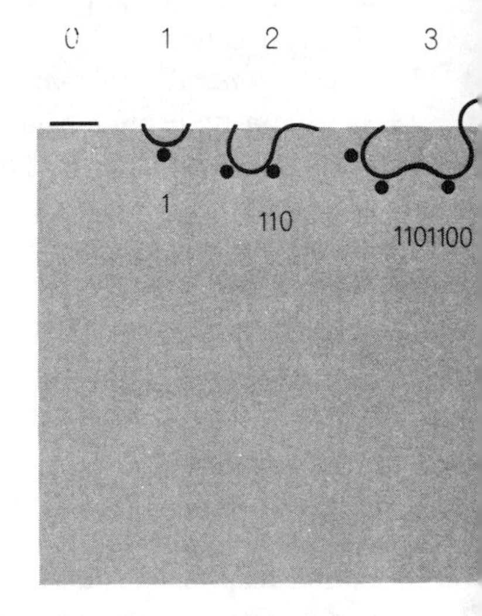

lowest order, but joined head to head so that the second is drawn from snout to tail.

Figure 104 shows dragon curves of orders 0 to 6. All dragons are drawn from tail to snout and are here turned so that each is swimming to the right, the tips of his snout and tail touching the waterline. If each 1 is taken as a symbol of a right turn instead of a left and each 0 as a left turn, the formula produces dragons that face the opposite way. The spots on each curve correspond to the central 1's in the formulas for the successive orders from 1 to the order of the curve. These spots, on a dragon of any order, lie on a logarithmic spiral.

The dragon curve was discovered by physicist John E. Heighway as the result of an entirely different procedure. Fold a sheet of paper in half, then open it so that the halves are at right angles and view the sheet from the edge. You will see an order-1 dragon. Fold the same sheet twice, always folding in the same direction, and open it so that every fold is a right angle. The sheet's opposite edges will have the shapes of order-2 dragons, each a mirror image of the other. Folding the paper in half three times generates an order-3 dragon, as illustrated in Figure 105. In general n folds produce an order-n dragon.

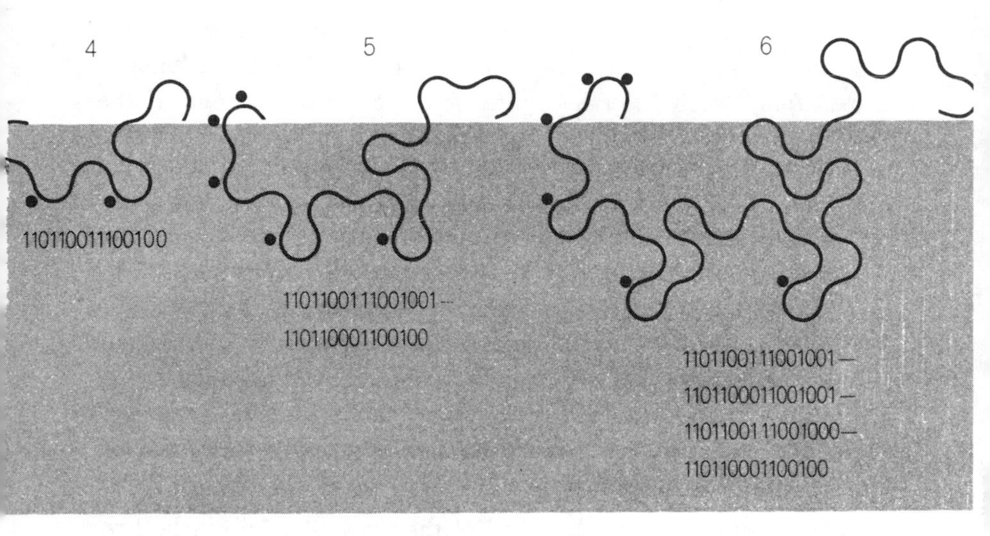

FIGURE 104
Sea dragons of orders 0 to 6, with their binary formulas

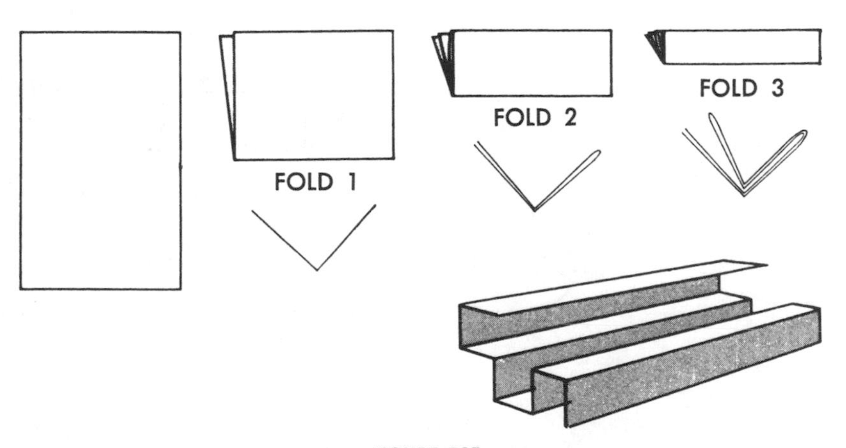

FIGURE 105
Three folds generate an order-3 dragon

The binary formula can be applied, of course, to the folding of a strip of paper (adding-machine tape works nicely) into models of higher-order dragons. Let each 1 stand for a "mountain fold," each 0 for a "valley fold." Start at one end of the strip, making the folds according to the formula. When the strip is opened until each fold is a right angle, it will have the shape of the dragon corresponding to the formula you used.

Physicist Bruce A. Banks discovered the geometric construction shown in Figure 106. It begins with a large right angle. Then at each step each line segment is replaced by a right angle of smaller segments in the manner illustrated. This is analogous to the construction of the "snowflake curve," as explained in my *Sixth Book of Mathematical Games from Scientific American*, Chapter 22. The reader should be able to see why this gives the same result as paper folding.

William G. Harter, the third of the three physicists who first analyzed the dragon curve, has found a variety of fantastic ways in which dragons can be fitted together snugly, like pieces

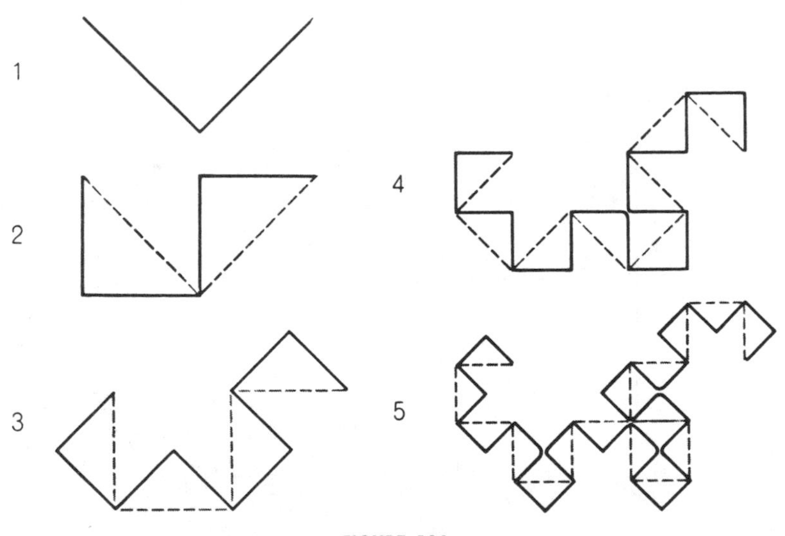

FIGURE 106
Geometric method

FIGURE 107
Four order-6 dragons joined at their tails

of a jigsaw puzzle, to cover the plane or to form symmetrical patterns. They can be joined snout to snout, tail to tail, snout to tail, back to back, back to abdomen, and so on. Figure 107 shows a tail-to-tail-to-tail-to-tail arrangement of four right-facing order-6 dragons. If the reader wishes to produce an eye-dazzling pattern, let him fit together in this way four order-12 dragons like the one shown in Figure 99. If the four curves are each infinite in length, they completely fill the plane in the sense that every unit edge of the lattice is traversed exactly once. For dragon-joining experiments it is best to draw your dragons on transparent paper that can be overlapped in various ways.

Donald E. Knuth, a Stanford University computer scientist, and Chandler Davis, a University of Toronto mathematician, have made the most extensive study of dragon curves. Their two-part article "Number Representation and Dragon Curves" (*Journal of Recreational Mathematics*, Vol. 3, April 1970, pages 66–81, and Vol. 3, July 1970, pages 133–49) is filled with material on ways of obtaining the dragon-curve sequence, variations and generalizations of the curve, and its properties. See also the article by Knuth and his wife, Jill, in the same journal, Vol. 6, Summer 1971, pages 165–67, which describes how they used three types of ceramic tiles to cover a wall of their house with an order-9 dragon.

8. If n soldiers of differing height stand in a row, at least p soldiers will be in either ascending or descending order. The number p is the square root of the smallest perfect square that is not less than n.

To prove this, label each soldier with a pair of letters, a and d. Let a be the maximum number of men on the soldier's left, including himself, who are in ascending height order. (By "left" I mean to your right as you face the row of soldiers.) Let d be the maximum number on his left, including himself, in descending order. It is easy to show (this is left to the reader) that no two soldiers can have the same pair of numbers. Their a numbers or their d numbers may be the same, but not both.

Assume that 10 soldiers are so arranged that their largest subset in ascending or descending order has p members, the lowest possible. No soldier can have an a or a d number greater than p. Since no two soldiers have identical pairs of a and d numbers, p must be large enough to provide at least 10 different pairs of a and d numbers.

Can p equal 3? No, because this provides only $3^2 = 9$ pairs of numbers:

a	1	1	1	2	2	2	3	3	3
d	1	2	3	1	2	3	1	2	3

Any number p will provide p^2 pairs of a and d numbers. Since 3^2 is 9, we do not have enough pairs to associate with the 10 soldiers. But 4^2 is 16, more than enough. We conclude that no matter how 10 soldiers arrange themselves, at least four must be in order. Four remains the p number for sets of soldiers up to and including 16. But 17 soldiers have a p number of 5 because we have to go to the next-highest p number to find enough a and d number pairs.

If 100 soldiers of different heights stand in a row, it is not possible for less than 10 to be in panpipe order. But add one more soldier and the p number jumps to 11.

A good discussion of the problem appears in Section 7 of "Combinatorial Analysis," by Gian-Carlo Rota, in *The Mathematical Sciences*, a collection of essays edited by the National Research Council's Committee on Support of Research in the Mathematical Sciences (M.I.T. Press, 1969). For generalizations and extensions of the problem see two articles: "Monotonic Subsequences," by J. B. Kruskal, Jr., in *Proceedings of the American Mathematical Society*, Vol. 4, 1953, pages 264–74, and "Longest Increasing and Decreasing Subsequences," by Craige Schensted, in *Canadian Journal of Mathematics*, Vol. 13, 1961, pages 179–91.

9. The fifth number is 0. The answer is, of course, trivial and intended as a joke. However, a difficult question now arises: Is there a fifth *positive* integer (other than 1, 3, 8, 120) which can be added to the set so that the set retains the property that the product of any two members is one less than a perfect square?

This unusually difficult Diophantine problem goes all the way back to Fermat and Euler. (See L. E. Dickson, *History of the Theory of Numbers,* Vol. 2, pages 517 f.) The problem has had an interesting history and was not finally settled until 1968. A student of C. J. Bouwkamp, at the Technological University, Eindhoven, Holland, saw the problem in *Scientific American,* mentioned it to Bouwkamp, who in turn gave it to his colleague J. H. van Lint. In 1968 van Lint showed that if 120 could be replaced by a positive integer, without destroying the set's property, the number would have to be more than 1,700,000 digits. Alan Baker, of Cambridge University, then combined van Lint's results with a very deep number theorem of his own, and finally laid the problem to rest. In a paper by Baker and D. Davenport, *Quarterly Journal of Mathematics,* second series, Vol. 78, 1969, pages 129–38, it is proved that there is no replacement for 120, and of course it follows that there can be no fifth member of the set. The proof is complicated, involving the calculation of several numbers to 1,040 decimal places.

It is known that there is an infinity of sets of four positive integers with the desired property, of which 1, 3, 8, and 120 has the smallest sum. You will find twenty other solutions listed in a discussion of the problem by Underwood Dudley and J. H. Hunter in *Journal of Recreational Mathematics,* Vol. 4, April 1971, pages 145–146. A simpler proof that there is no fifth number for the 1, 3, 8, 120 set is given by P. Kanagasabapathy and Th. Ponnudurai in a paper published in *The Quarterly Journal of Mathematics,* Vol. 3, No. 26, 1975, pages 275–278. The problem is also the topic of "A Problem of Fermat and the Fibonacci Sequence," by V. E. Hoggatt, Jr., and G. E. Bergum, in *The Fibonacci Quarterly,* Vol. 15, December 1977, pages 323–330.

Is there a set of five positive integers with the desired property? As far as I know, this remains unanswered.

CHAPTER 16

Colored Triangles and Cubes

In 1967 Franz O. Armbruster, a California computer programmer, redesigned a tantalizing little puzzle that has been marketed in dozens of different forms for more than half a century. He packaged it cleverly and inexpensively with brief, witty instructions and called it Instant Insanity. It was an instant success. Parker Brothers took it over and in 1968 its sales were astonishing. The puzzle consists of nothing more than four plastic cubes, all the same size, each face bearing one of four colors. The problem is simply to arrange the cubes in a straight row so that all four colors appear on each of the row's four sides.

I had mentioned this puzzle in a chapter, "The 24 Color Squares and the 30 Color Cubes," in *New Mathematical Diversions from Scientific American* (1966), but the puzzle's most complete analysis is to be found in Chapter 7 of *Puzzles and Paradoxes* (1965) by the Glasgow mathematician Thomas H. O'Beirne. O'Beirne calculated the probability of solving the puzzle by chance as one in 41,472 random tries! He wrote that the most tantalizing feature of the Tantalizer, as it was called in one of its most recent incarnations, is that it "can be brought out again and again, with trivial variations, while many other good puzzles appear once and vanish, or circulate only privately, if at all!"

Instant Insanity can be considered one of a huge general class of combinatorial problems in which regular polygons or polyhedrons with their edges or faces colored or distinguished by numbers or other symbols are to be fitted together under certain restraints to achieve specified results. One of the pioneering authorities on combinatorial mathematics, Major Percy Alexander MacMahon, who died in 1929, devoted a great deal of thought to such puzzles. MacMahon, a professor of physics and a mathematician, was the author of the classic two-volume *Combinatory Analysis* (1915, 1916) and an excellent introductory article on the same topic for the eleventh edition of the *Encyclopaedia Britannica*. He also wrote a little-known and long-out-of-print book called *New Mathematical Pastimes* (1921), in which he explored a large variety of puzzles of the general type characterized here.

In my chapter on the 30 color cubes (a remarkable set of

FIGURE 108
The 24 color triangles

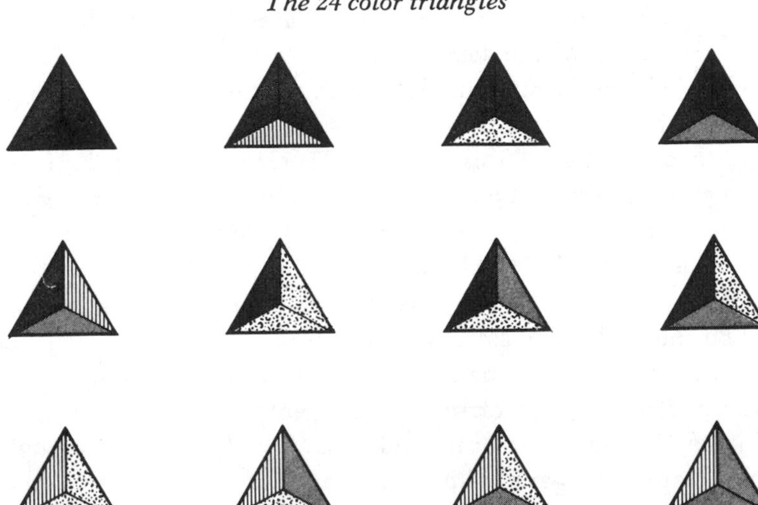

cubes discussed by MacMahon in *New Mathematical Pastimes*) I also dealt with MacMahon's two-dimensional set of 24 color squares. This chapter introduces MacMahon's companion set of 24 color triangles. If the three edges of an equilateral triangle are each colored with one of two colors, and if rotations of triangles are not considered different, a set of four triangles results. Three colors produce a set of 11 distinct triangles and four colors the set of 24 triangles shown in Figure 108. To work with such a set, cut from cardboard, it is convenient to divide each triangle into three identical triangular parts as shown and then color each part, using four contrasting colors for the four differently labeled regions. Because triangles are not to be turned over (the set includes mirror-image pairs), only one side of the cardboard should be colored.

The formula for the number of different equilateral triangles that can be produced in this way, given n colors, is

$$\frac{n^3 + 2n}{3}.$$

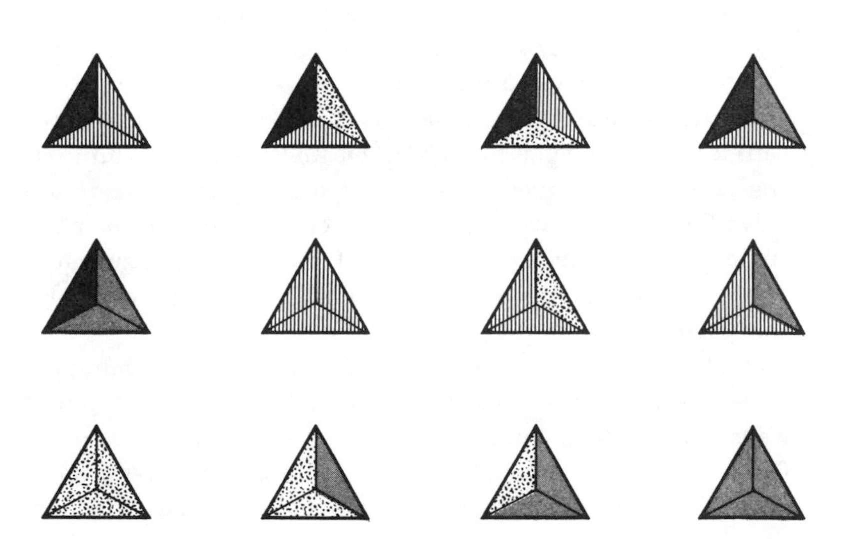

When $n = 3$, the resulting set of 11 triangles is small and will not form any interesting shapes. When $n = 5$, the set of 45 triangles is a little too large for recreational purposes. The set of 24, using four colors, is just about right; moreover, its pieces will form a regular hexagon as well as an unusually large number of different symmetrical shapes. MacMahon gives many combinatorial problems connected with this set. The simplest problems treat the pieces as triangular "dominoes" to be fitted together with adjacent edges matching in color to form symmetrical polygons. To this he adds a second restriction: the entire border of the polygon must be the same color. (Henceforth these will be referred to as MacMahon's two provisos.) Since each color appears in the set on 18 edges (an even number) and since the contact proviso requires that the color appear within a polygon on an even number of edges, it follows that the perimeter of any shape solvable under the two provisos must have a border composed of an even number of edges.

Wade E. Philpott, a retired engineer who lives in Lima, Ohio, has done more work on this set of MacMahon color triangles than anyone else I know of. What follows is taken from my correspondence with him and is presented here with his permission.

It is not hard to prove that all polygons formed by the 24 color triangles under MacMahon's two provisos must have perimeters of 12, 14, or 16 unit edges. The perimeter, as we have seen, must have an even number of edges. The minimum-length perimeter of a polygon formed by putting together 24 unit triangles is 12. A perimeter of 18 is impossible because there are only 18 edges of a given color in the set and the solid-color triangle cannot contribute all three of its edges to a polygon. Therefore 16 is the maximum length of the perimeter of a solvable polygon.

Only one polygon, the regular hexagon, has the minimum perimeter of 12. Its one-color border can be formed in six different ways, each with an unknown number of different solutions. Philpott estimates the total number of solutions as several thousand. (This does not include rotations and reflections as different, or new solutions obtained simply by interchanging colors.) Philpott has discovered that for each type of border the hexagon can be solved with the three triangles of solid color (necessarily differing in color from the border) placed symmetrically around the center of the hexagon. Since each solid-color triangle must be surrounded by triangular segments of the same color, the result is three smaller regular hexagons of solid color situated symmetrically within the larger hexagon. Figure 109 shows the general schemata for the six different ways of making the one-color border, with the solid-color triangles symmetrically placed at the hexagon's center in one of two possible ways. The reader may enjoy trying to construct a hexagon for each of the six patterns.

The 24 triangles will form two kinds of parallelogram: 2 × 6 and 3 × 4. It is easy to prove that the 2 × 6 cannot meet MacMahon's two provisos: the parallelogram has 14 triangles that contribute an edge to its perimeter, but only 13 triangles bear the same color. The 3 × 4 parallelogram *is* solvable. Again, the total number of solutions is not known, although Philpott guesses it is less than for the regular hexagon. Like the hexagon, it has six different types of border. Figure 110 shows a solution by Philpott for each type, each with the three solid-color triangles (necessarily differing from the border color) arranged in a row.

The 3 × 4 parallelogram is an example of a symmetrical shape with a perimeter of 14 unit edges. Philpott has found 18 polygons with 14-edge perimeters that possess symmetry, either reflectional or rotational, or both, and that are solvable under MacMahon's two provisos. The 18 are reproduced in Figure 111. All have more than one solution. It is easy to see that to be solvable a 14-edge polygon must have at least one "point" (a 60-degree corner) because at least one triangle with adjacent edges of the same color must contribute both of those edges to the perimeter. Note that only one of the 18 figures (the first one) has a single point. This is also one of 11 symmetrical shapes of 14 or 16 edges for which there is only one type of one-color perimeter. Shape 5 is also of special interest. Accord-

FIGURE 109

The six types of border for a regular hexagon

FIGURE 110

A solution for each type of border for the 3 × 4 parallelogram

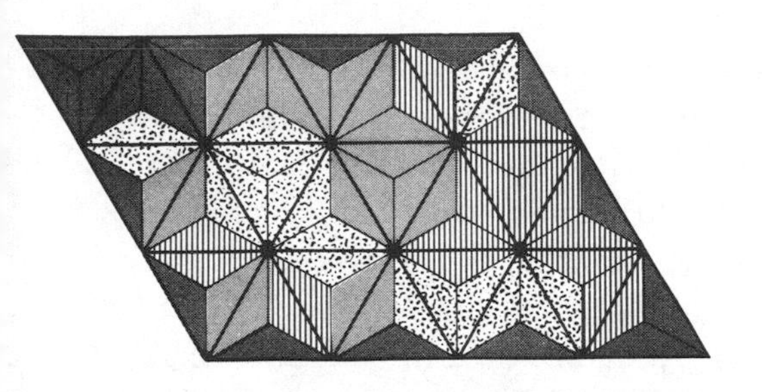

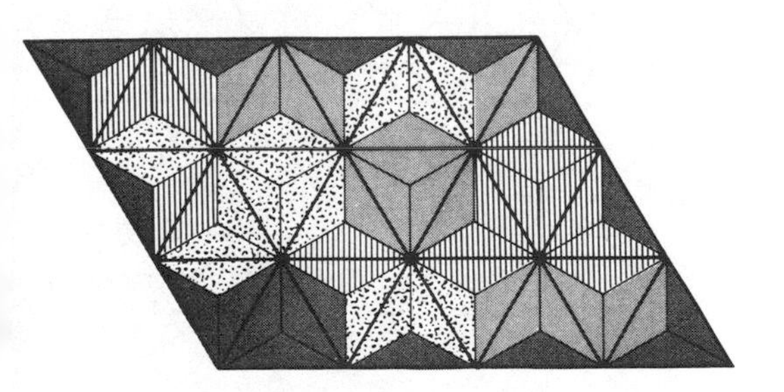

ing to Philpott, it is the only symmetrical pattern that has 11 kinds of one-color border (the maximum possible). Solutions are known for each of these 11 forms.

Philpott found 42 solvable symmetrical shapes with 16-edge perimeters [*see Figure 112*], making a total of 61 solvable symmetrical polygons in all. He reports that all solvable 16-edge polygons must have at least three points and not more than four. Not all three-point symmetrical shapes are solvable, but all four-point shapes are.

The "duplication problem," proposed by Philpott, is to form two identical symmetrical shapes, of 12 triangles each, that meet MacMahon's two provisos and have borders of solid colors which necessarily differ. Figure 113 shows one of the 28 known solvable shapes. Philpott estimates the number of different solutions for each shape to be in the hundreds, not thousands.

Philpott's "triplication problem" is to use the 24 triangles to form three identical symmetrical shapes, of eight triangles each, with perimeters of three colors. It has been proved, he reports,

FIGURE 111
The 18 solvable symmetrical polygons with 14-edge perimeters

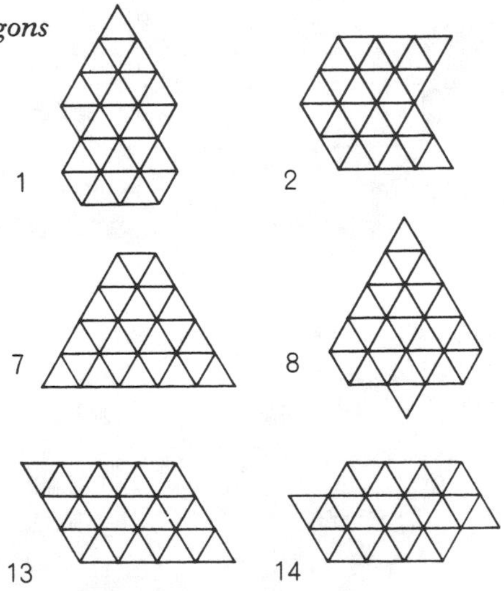

that there are just ten such shapes. They are reproduced in Figure 114 with a solution for one of them. Philpott estimates the number of solutions for each to be less than 100.

Six triangles will form 12 different shapes, not all symmetrical, that are known as hexiamonds. (See Chapter 18 of my *Sixth Book of Mathematical Games from Scientific American.*) All hexiamonds except the "butterfly," Philpott has found, can be quadrupled.

John Harris, of Santa Barbara, California, suggested the problem of constructing hexagons with a minimum or a maximum number of isolated "diamonds." (A diamond is formed by two tiles with meeting edges of the same color.) It is easy to show that there must be at least one such diamond and not more than nine. Solutions, not unique, exist for both cases. Nine diamonds, Harris found, are possible on the 3×4 parallelogram, with many solutions. The parallelogram may be formed with *no* diamonds, again with many solutions.

What complete sets of MacMahon triangles will form an

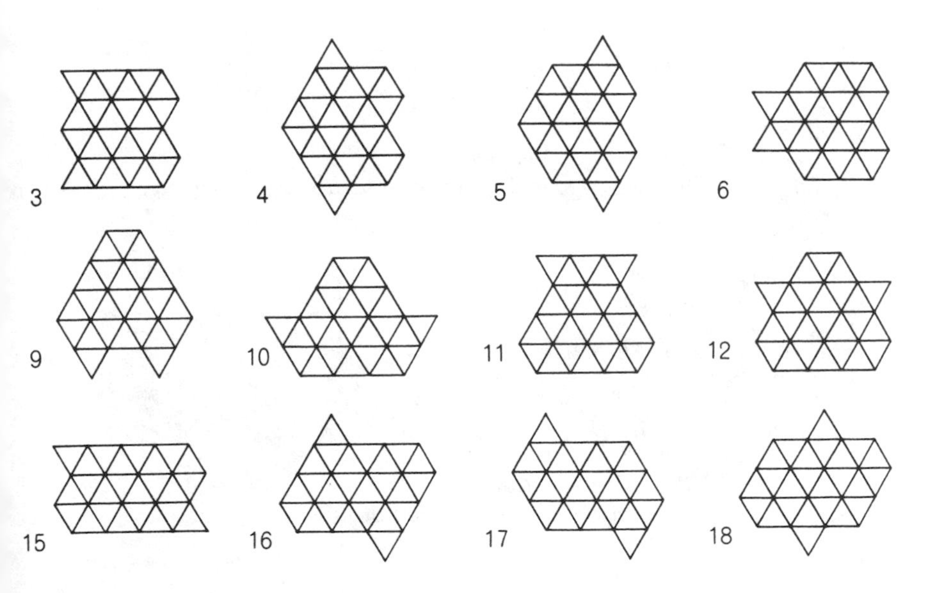

FIGURE 112

The 42 solvable symmetrical polygons with 16-edge perimeters

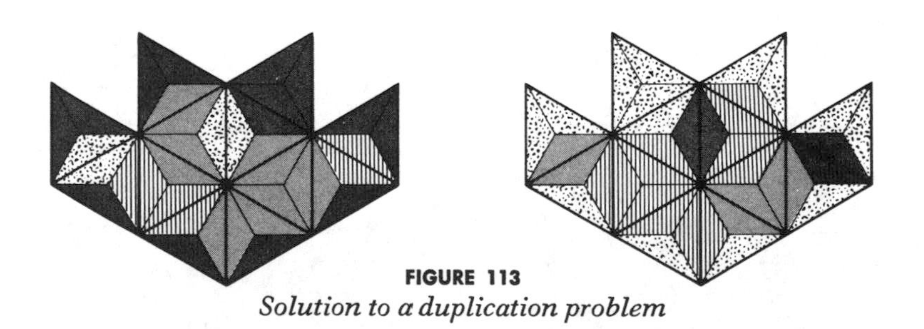

FIGURE 113

Solution to a duplication problem

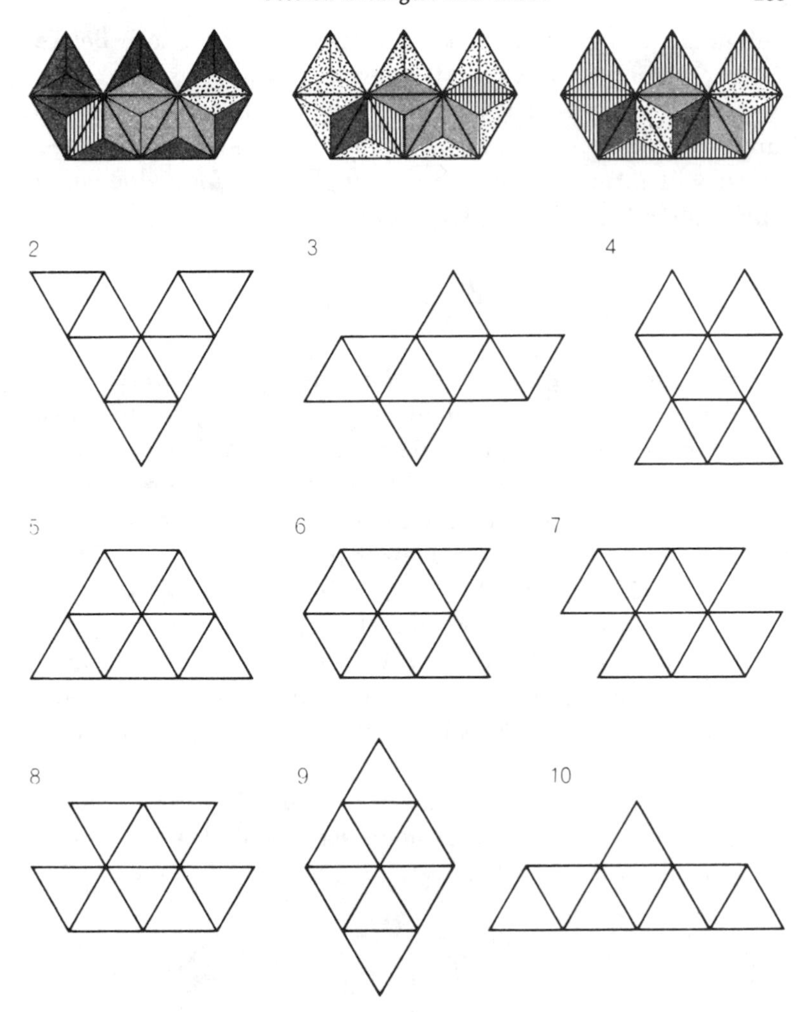

FIGURE 114
The 10 symmetrical polygons that can be triplicated, one of them solved

equilateral triangle that meets the two color provisos? Before answering this we must first determine what sets will form equilateral triangles *without* the provisos. Letting n be the number of colors in a complete set, and m^2 the number of tiles, the set will form an equilateral triangle only for values of n which satisfy the following Diophantine equation:

$$\frac{n\,(n^2+2)}{3} = m^2.$$

How many integral solutions does the equation have? Philpott published this as a problem in the *Journal of Recreational Mathematics*, Vol. 4, April 1971, page 137. A partial answer appeared in Vol. 5, January 1972, pages 72–73. There is a finite number of solutions of which the smallest are $n = 1$, 2, and 24, and no other solutions for n less than 5,000.

A single triangle ($n = 1$) meets both color provisos trivially, and when $n = 2$ it is obvious that the complete set of $m^2 = 4$ tiles will not meet the border proviso. When $n = 24$, the $m^2 =$ 4,624 tiles form a triangle of 68 units on the side. Will they also meet the two provisos? Probably, but this has not yet been demonstrated.

George Littlewood, of Manchester, England, proved that the triangles in a complete MacMahon set would make a regular hexagon only when $n = 4$. This follows from the fact that

$$\frac{n\,(n^2+2)}{3} = 6m^2$$

has an integral solution only when $n = 4$. As we have seen, it is possible to form such hexagons and meet the two color provisos. Excluding rotations, reflections, and color permutations, how many such hexagons are there? This is not yet established. Philpott estimates the number to be several thousand.

A set of 45 tiles for the set of five-color triangles (edges marked with zero, one, two, three, or four black spots instead of colors) was marketed in West Germany in the late 1960's under the name of Trimino. The boxed set included a booklet by Heinz Haber giving symmetrical shapes to be made by the pieces and instructions for playing a competitive game. The set includes, of course, the 24 four-color triangles as a subset. A similar set was imported into the United States from Hong Kong as a game called Three Dimensional Dominoes. Although sets of the 24 four-color triangles alone have surely been marketed from time to time, the first such set I can document is a magnetized pocket set made in London by Just Games, Ltd., that I saw advertised in 1975.

In 1892 MacMahon obtained British patent No. 3,927 on his set of 24 four-color triangles, but I do not know if they were ever marketed. In the United States, in 1895, a set of four-color, edge-colored triangles was patented (No. 331,652) by F. H. Richards, of Troy, New York. However, Richards describes their use only in playing domino-type games. Various domino games using edge-colored triangles have been sold in this country, notably Contack (Parker Brothers, 1939) and Al-lo-co, put out by a Cleveland firm in 1964.

Instead of coloring edges, each colored edge can be replaced by one of four types of symmetrically shaped edge. This was proposed by MacMahon as a way of transforming his color triangles into equivalent jigsaw puzzles. [*See Figure 115.*]

MacMahon's book does not consider triangles colored at their corners instead of edges. The number of such triangles, for each value of n, is the same as when they are edge-colored. Will a set of 24 four-color, corner-colored triangles form a hexagon with the single proviso that corners of like colors meet at every vertex? Unfortunately, no. Nor is it possible to make an equilateral triangle with a symmetrically placed triangular "hole," although it can be made if the vacancy is at a corner or the center of a side. The set will, however, make a 3×4 parallelogram as well as many other symmetrical shapes.

In 1969 Marc Odier, of Paris, designed a set of 24 four-color, corner-colored triangles which were made and sold in France as the game of Trioker. The game is protected by British patent No. 1,219,551. It includes a set of patterns to be solved, and instructions for a competitive game. A 25th tile of two colors,

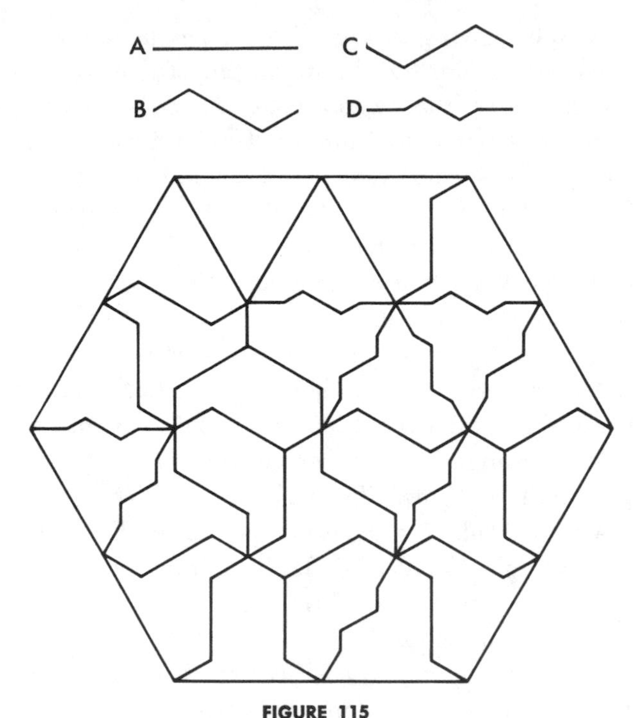

FIGURE 115
Jigsaw puzzle using triangles made with four different edges

called the Joker, is sometimes used for both patterns and the game. More recently Trioker has been marketed in Spain. In 1976 a 207-page book on recreations with this set of tiles, *Surprenants Triangles*, by Odier and Y. Roussel, was published in France by CEDIC. A game using four-color, corner-colored triangles called Tri-ominoes, made by Pressman, was on sale in the United States in 1968.

In three dimensions the cube is the only regular solid replicas of which will fit together to fill space. This surely is one reason for its being used in so many different combinatorial puzzles of the Instant Insanity type. If the reader will obtain 27 identical cubes (alphabet blocks will do) and paint 9 one color (on all sides), 9 another color, and 9 a third color, he will have the material for working on two unusual three-dimensional combinatorial problems.

It is obviously impossible to form the 27 cubes into a single $3 \times 3 \times 3$ cube so that each of its 27 orthogonal rows (rows parallel to an edge of the large cube) consists of three cubes of the same color. Can they form a cube in which all three colors appear on each of the 27 orthogonal rows? They can; a unique solution (not counting rotations, reflections, or permutations of colors as different) was discovered by Charles W. Trigg, a retired California mathematician. Can the reader rediscover it?

The second problem, much more difficult, was recently invented by the Cambridge mathematician John Horton Conway. Conway set himself the task of forming the $3 \times 3 \times 3$ cube so that *every* row of three (the cube's 27 orthogonal rows, its 18 diagonal rows on the nine square cross-sections, and its four space diagonals that join opposite corners) contains *neither* three cubes of like color nor three of three different colors. In other words, each of the 49 straight rows of three cubes will consist of two cubes of one color and one of another. Conway found two distinct but closely related solutions (again not counting rotations, reflections, or permutations of colors).

One can, of course, work on both problems by drawing three ticktacktoe boards to represent the three levels of the large cube and labeling the 27 cells properly with nine A's, nine B's, and nine C's. It is easier and more fun to work with actual cubes, however, and well worth the trouble of acquiring and coloring a set, even if it is only a set of crayoned sugar cubes.

ANSWERS

FIGURE 116 shows one way of forming the hexagon with the 24 color triangles for each of the six possible varieties of one-color border, with the added proviso that the three solid-color triangles be placed symmetrically around the center in the manner shown. It is not known how many solutions there are for each of these six types.

Figure 117 gives the unique solution (not counting rotations, reflections, or permutations of colors as different) for forming a $3 \times 3 \times 3$ cube with 27 unit cubes, nine of each of three colors,

FIGURE 116
Six solutions to the hexagon problem

so that each of the 27 orthogonal rows contains a cube of each color. The solution was published by Charles W. Trigg in *Mathematics Magazine,* January 1966.

Figure 118 shows the only two ways, both found by John Horton Conway, of arranging the same set of 27 cubes into a $3 \times 3 \times 3$ cube so that each of its 49 straight rows of three (orthogonal and diagonal, including the cube's four space diagonals connecting opposite corners) contains neither three cubes of the same color nor three cubes of three different colors.

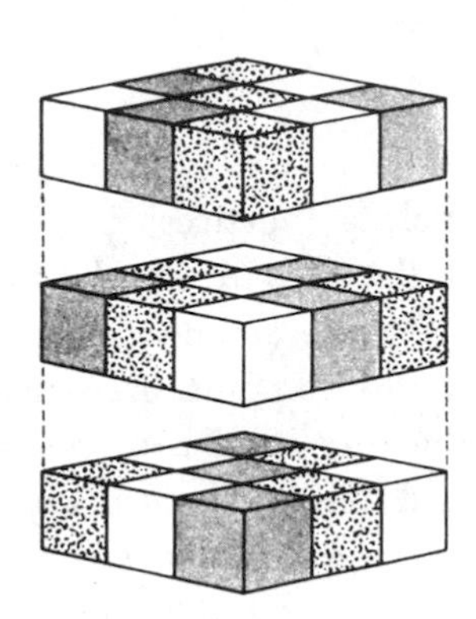

FIGURE 117
Solution to first cube problem

FIGURE 118
Two solutions to the second cube problem

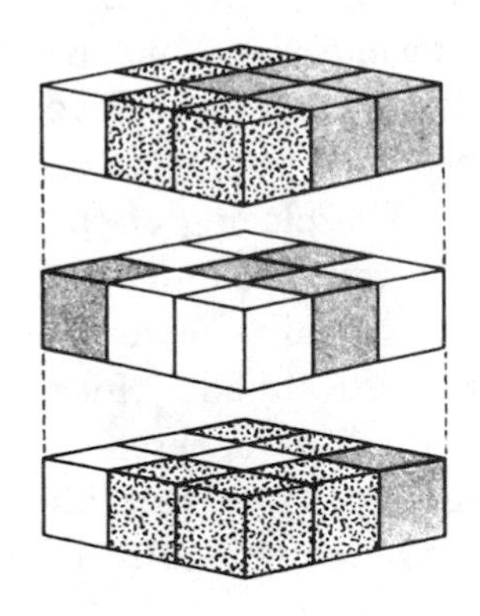

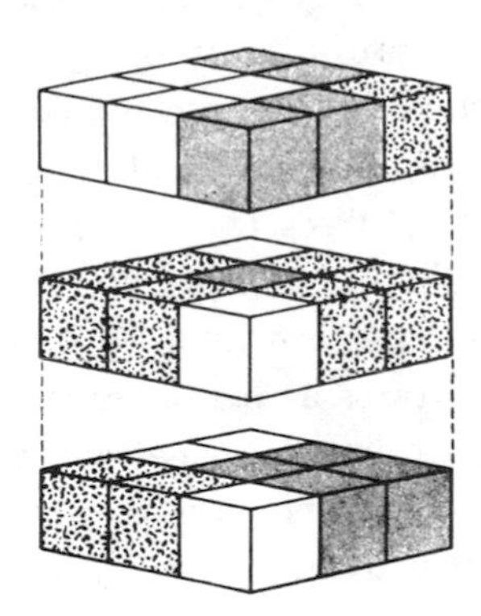

CHAPTER 17

Trees

A "CONNECTED GRAPH" is a set of points (vertices) joined by line segments (edges) in such a way that a path can be found from any point to any other point. If there are no circuits, or paths leading from a point back to the same point, a connected graph is called a "tree." In nature the tree itself is of course a splendid three-dimensional model, and there are also crystals that grow in a similar manner. Rivers and their tributaries sprawl over the earth's surface in gigantic tree diagrams. Certain brittle solids crack in such a way that the break, examined under a microscope, shows a beautiful treelike pattern. Electric discharges sometimes branch like trees.

The simplest tree graph is a line connecting two points. Three points also join in only one way to form a tree, but four points can be connected in two topologically distinct trees. Five points yield a "forest," or set, of three trees and six points can be connected in six trees [*see Figure 119*]. The placing of the points and the shapes of the edges are irrelevant because only topological properties are used here as distinguishing features; think of the diagrams as being formed of identical balls joined by elastic bands. These are called "free trees" as opposed to "rooted trees," in which one point is distinguished from all others, or "labeled trees," in which all points are distinguished.

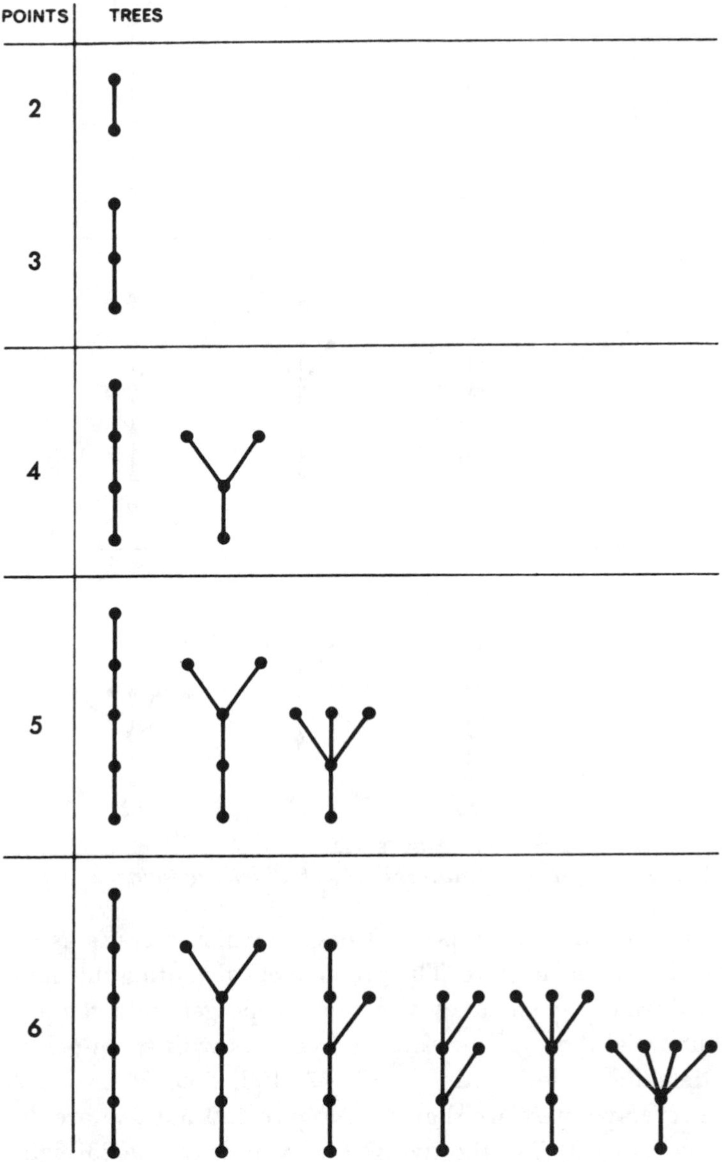

POINTS | TREES
2
3
4
5
6

FIGURE 119
Topologically distinct trees of two to six points

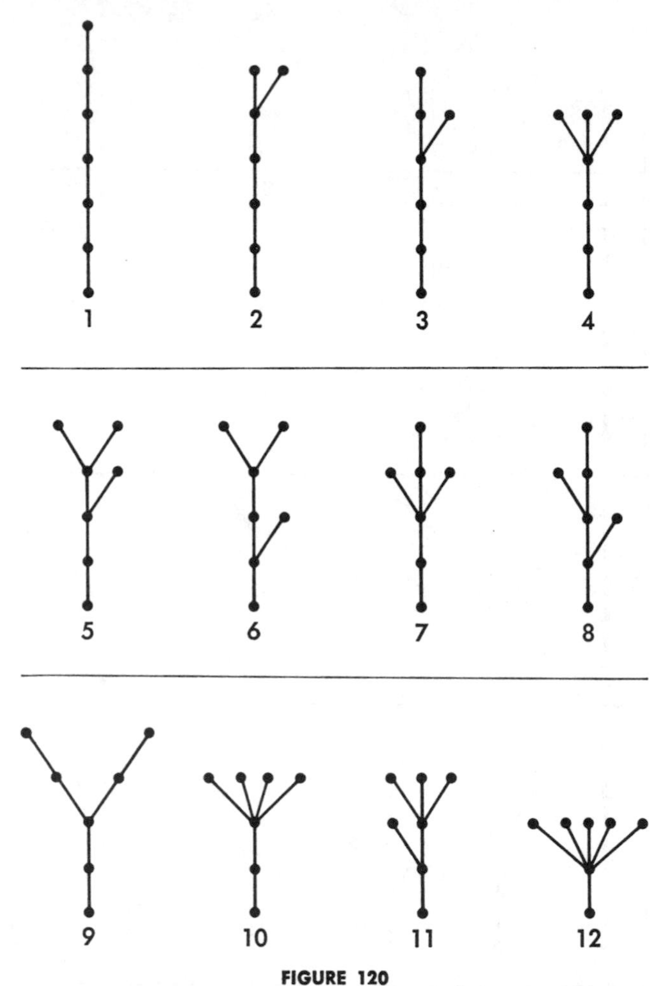

FIGURE 120

Twelve seven-point trees, two of which are twins

There are still other types of trees for which there is as yet no standard nomenclature. The problem of calculating the number of distinct n-point trees of a given type gets into complex combinatorial theory. There are 11 free trees with seven points, and then the series continues 23, 47, 106, 235, 551, . . . A dozen seven-pointers are shown in Figure 120 but two are duplicates. Can you find the twins? Can you draw the 23 eight-pointers?

It is obvious that every tree of n points has $n - 1$ edges, and a forest of n points and k trees has $n - k$ edges. Another obvious theorem is illustrated by a scene in L. Frank Baum's fantasy *The Magical Monarch of Mo and His People*. An apple on a high branch cannot be reached by climbing the tree because someone has sawed off part of the trunk, near the branch, to use for kindling. The theorem: Removal of any edge from a tree graph disconnects the graph. Even a terminal edge, if it is removed, leaves its terminal point stranded.

Investigations of the properties of tree graphs did not get under way until the late nineteenth century, but of course the diagram was in use in ancient times. It is a handy way to show all kinds of relations—genealogical ones, for instance—and for dividing a subject matter into hierarchic categories. One of the most ubiquitous tree graphs in medieval metaphysics was proposed in a commentary on Aristotle by Porphyry, a third-century Neoplatonist and opponent of Christianity. In essence the Tree of Porphyry is what is now called a "binary tree." Categories are split into two mutually exclusive and exhaustive parts on the basis of a property possessed by one part but not the other. (See Plato's *Phaedrus*.) Substance, the *summum genus*, divides into the corporeal and the incorporeal, the corporeal into the living and the nonliving. The living divides into the sensible (animals) and the insensible (plants). Animals divide into the rational (man) and the nonrational; the rational in turn splits into individual persons, the *infama species* of the tree. After the invention of engraving Renaissance philosophers liked to publish fantastically branched and elaborately decorated diagrams of the Porphyrian tree.

Petrus Ramus, the French Protestant logician killed in 1572 in the Massacre of St. Bartholomew's Day, was obsessed by this kind of exhaustive division and applied the binary tree to so many topics that it was thereafter known as the Tree of Ramus. Jeremy Bentham, in the early nineteenth century, was perhaps the last important philosopher to take the binary tree quite so seriously. Although he realized that a complete Ramean tree was unwieldy in many areas (for example, botany!) and that, like an apple, a category can often be halved in thousands of different ways, he was convinced that dichotomous division was one of the great tools of analysis. He wrote of the "matchless beauty of the Ramean tree," and headed a section of one essay "How to plant a Ramean encyclopedical tree on any given part of the field of art and science."

Philosophers today (aside from those working in formal logic) have little use for tree diagrams, but mathematicians and scientists have found applications for them in such diverse fields as chemical structure, electrical networks, probability theory, biological evolution, operations research, game strategy, and all kinds of combinatorial problems. The most striking example I know of the unexpected applicability of tree diagrams to a combinatorial problem (in this case a game of card solitaire) is given in a discussion of tree theory in Donald E. Knuth's book *Fundamental Algorithms.*

The solitaire game is best known as "clock," although it also goes by such names as "travelers," "hidden cards," and "four of a kind." The pack is dealt into thirteen face-down piles of four cards each, the piles arranged as shown in Figure 121, left, to correspond to the numbers on a clock face. The thirteenth

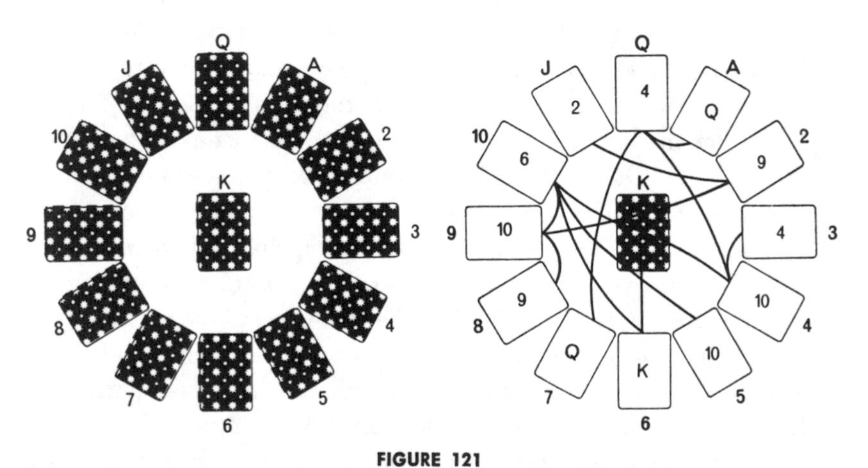

FIGURE 121

Starting position for clock solitaire

Bottom cards and their tree connections

(king) pile goes in the center. Turn over the top card of the king pile, then slide it face up under whichever pile corresponds to the card's value. For example, if it is a four, put it under the four o'clock pile; if a jack, under the eleven o'clock pile, and so on. Now turn up the top card of the pile under which you just placed the card and do the same thing with the new card. The play continues in this way. If you turn a card that matches the pile it is in, slide it face up under that pile and turn the next top card. The game is won if you get all fifty-two cards face up. If you turn a fourth king before this happens, the play is blocked and the game lost.

Playing clock is purely mechanical, demanding no skill. Knuth proves in his book that the chances of winning are exactly 1/13 and that in the long run the average number of cards turned up per game is 42.4. It is the only known card game, given in popular books on solitaire, for which the probability of winning has been precisely calculated.

Knuth also discovered a simple way to know in advance, merely by checking the bottom card of each pile, whether the game will be won or lost. Draw another clock-face diagram, but this time indicate on each pile the value of the bottom card of that pile—except for the center, or king, pile, the bottom card of which remains unknown. Now draw a line from each of the twelve bottom-card values to the pile with the corresponding number [*see Figure 121*, *right*]. (No line is drawn if the card's value matches its own pile.) Redraw the resulting graph to reveal its tree structure [*see Figure 122*]. If and only if the graph is a tree that includes all thirteen piles will the game be won. The arrangement of the forty unknown cards is immaterial.

The illustrated game, as the tree graph reveals, will be won. The reader is invited to draw a similar diagram for another starting position [*see Figure 123*] to determine whether it is a win or loss, and then to check the result by actually playing the game. A proof that the tree test always works will be found in Knuth's book. In addition to being the introductory volume of what will surely be a monumental survey of computer science,

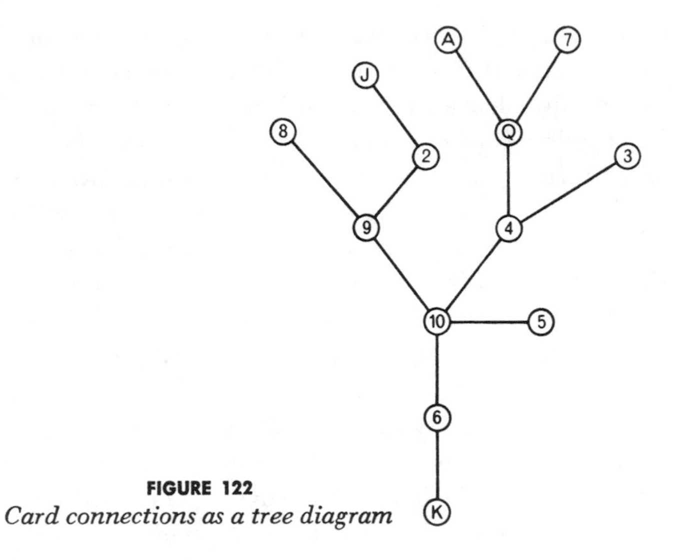

FIGURE 122
Card connections as a tree diagram

it is crammed with fresh material that is of great interest to recreational mathematicians.

A tree that catches all of a set of points is said to be a "spanning tree" for those points. One of the earliest theorems in tree theory was the discovery, by the nineteenth-century Cambridge mathematician Arthur Cayley, that the number of different spanning trees for n labeled points is n raised to the power of $n - 2$. (Cayley was one of the founders of tree theory, which he developed in 1875 as a method of calculating the number of different hydrocarbon isomers.) Suppose there are four towns, A, B, C, and D. If we join them with a spanning tree, in how many different ways can it be done? Cayley's formula gives 4^2, or 16 [*see Figure 124*]. There are topological duplications, but because the vertices (towns) are distinguished we count each as being different. Where crossings occur one edge is shown going under the other to make it clear that the crossing is not another vertex; otherwise the tree would be a five-point tree.

Suppose n towns are to be joined by a railroad network consisting of straight track segments connecting pairs of towns. Tracks may cross, but if so the crossings must not be taken as new vertices; that is, they are not points at which a traveler can

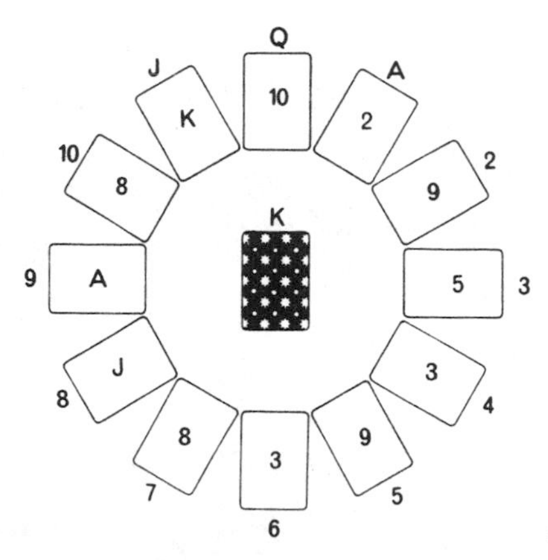

FIGURE 123
Bottom cards to be treed

transfer from one track to another. By what procedure can you find a spanning graph that has the smallest total length?

It is easy to see that the minimum graph is a tree. Otherwise it would contain a circuit, in which case the graph's length could be made shorter by the removal of one edge, breaking the circuit but leaving all towns still connected. Since any circuit can be eliminated, shortening the graph, the minimum graph will be a tree.

There are several simple algorithms (procedures) for finding a minimum-length spanning tree. The standard procedure (first given by Joseph B. Kruskal, "On the Shortest Spanning Subtree of a Graph and the Traveling Salesman Problem," *Proceedings of the American Mathematical Society*, Vol. 7, February 1956, pages 48–50) is as follows. Determine the distance between each pair of towns, then label these distances in increasing order of length. The shortest is 1, the next shortest 2, and so on. If two distances are equal it does not matter which is numbered first. Draw a straight line between the two towns separated by distance 1. Follow with similar straight lines between pairs of towns with distances 2, 3, 4, and so on. Never add an edge that completes a circuit. If drawing a line produces a circuit, ignore that pair of towns and go on to the next-higher

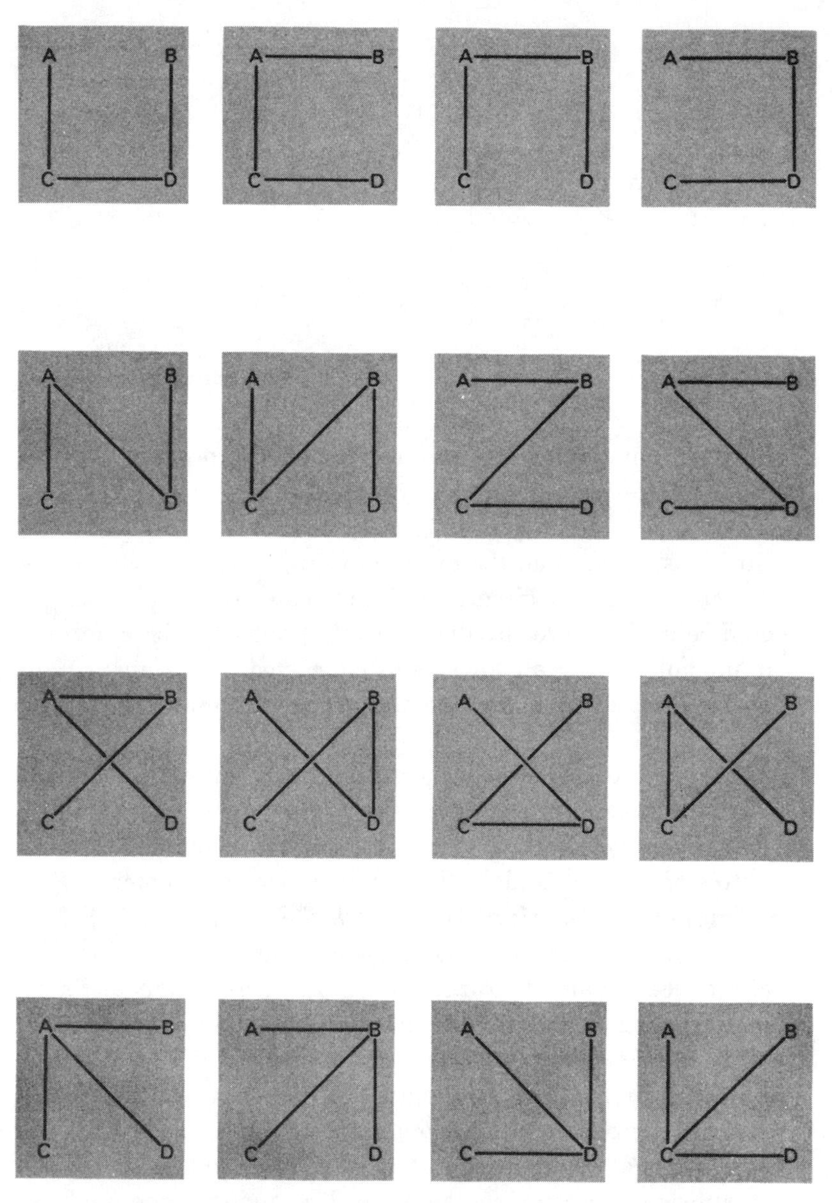

FIGURE 124
The 16 labeled trees that span four points

distance. The final result is a spanning tree of minimum length. There may be other spanning trees of the same length but "Kruskal's algorithm" is sure to construct one of them.

Minimum spanning trees have many interesting properties that are not difficult to prove. The edges intersect only at vertices, for example, and no vertex need have more than five edges meeting it.

The "economy tree problem," as this is sometimes called, should not be confused with the "traveling salesman problem," a famous unsolved problem of graph theory. In that problem one seeks the shortest circuit enabling a salesman to visit every town once and only once and return to his starting town. There are good computer algorithms for finding close approximations to the shortest circuit when the number of towns is large, but there is no absolutely accurate general procedure except the tedious testing of all possible routes.

If one is allowed to join towns by trees that contain new vertices, then the shortest tree is called a Steiner tree. For example, what is the shortest railroad connecting four towns at the corners of a square? Assume that the square's side is one mile. Remember, the minimal spanning tree in this case may contain one or more additional vertices; it need not be a four-point tree. If the reader succeeds in finding this tree, he can try the more difficult problem of determining the minimal-length Steiner tree joining the five corners of a regular pentagon.

ANSWERS

In the illustration of free trees with seven points, the two duplicates are trees 5 and 8. The second starting position for clock solitaire is a loss. Its graph is not a tree; not only is it disconnected but also one part contains a circuit.

Figure 125 shows how to draw minimal-length Steiner trees joining the corners of a square and of a regular pentagon. The dotted angles are 120 degrees. It might be thought that the two diagonals of a square would provide an "economy tree" (length $2\sqrt{2} = 2.828+$) connecting a unit square's corners, but the net-

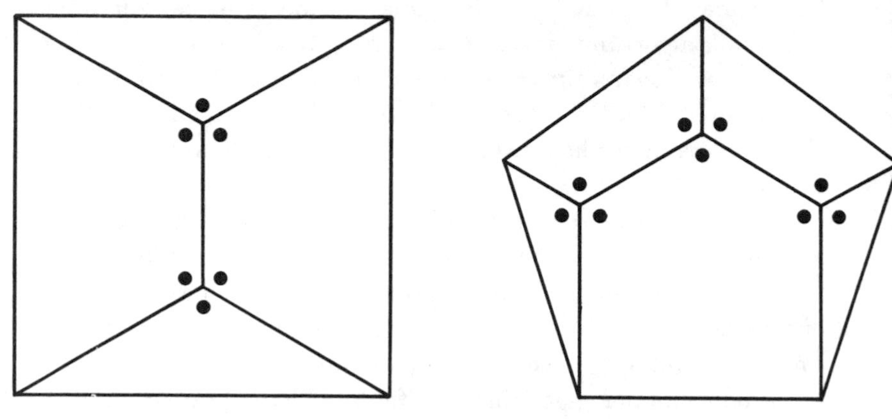

FIGURE 125

Economy trees connecting corners of square and pentagon

work shown has a length of $1 + \sqrt{3} = 2.732+$. A proof without calculus that this is minimal is given in Hugo Steinhaus' *100 Problems in Elementary Mathematics*, problem 73 (Basic Books, 1964). The minimal path inside a pentagon of unit side is 3.891+.

The minimum Steiner tree inside an equilateral triangle has a fourth point in the triangle's center. Minimum Steiner trees on regular polygons of six or more sides are simply the perimeter with one side removed. For the general problem of finding minimal Steiner networks connecting n points on the plane—and a technique for finding such networks by capitalizing on the surface tension of a film of soap solution—see Chapter 7 of *What Is Mathematics?* by Richard Courant and Herbert E. Robbins (Oxford, 1941).

CHAPTER 18

Dice

We figured the odds as best we could, and then we rolled the dice.

—JIMMY CARTER,
quoted by *The New York Times*, June 10, 1976,
on his decision to seek the Presidency

THE CHANCE ELEMENT in thousands of indoor games is introduced by a variety of simple random-number generators. The most popular of such devices, ever since the time of ancient Egypt, have been cubical dice. Why cubical? Because of their symmetry, any of the five regular solids can be and have been used as gaming dice, but the cube has certain obvious advantages over the other four solids. It is the easiest to make, its six sides accommodate a set of numbers neither too large nor too small, and it rolls easily enough but not too easily.

The four-sided tetrahedron has been the least popular over the centuries; it hardly rolls at all and will randomize no more than four numbers. Next to the cube the octahedron has been most used as a randomizer for games. Specimens of octahedral dice have been found in ancient Egyptian tombs and are still used today in certain games. Dodecahedrons (twelve sides) and icosahedrons (twenty sides) have been employed mainly for fortune-telling. A dodecahedral die was a popular fortune-telling device in sixteenth-century France, and if you break open one of those large fortune-telling balls in which answers to questions float upward in a liquid and appear in a window at the top, you will find the answers printed on the twenty faces of a floating icosahedron.

A few years ago the Japanese Standards Association (Kobiki-kan-Bekkan Building, 6–1, Ginza-higashi, Chuo-ku, Tokyo) found a practical use for icosahedral dice. Since the number of an icosahedron's sides is twice 10, pairs of its sides can be marked with one of the ten digits from 0 through 9 to make an elegant little instrument for generating random decimal digits to be used in Monte Carlo methods, game theory, and so on. The dice are sold in sets of three, each a different color (red, blue, yellow) so that every throw produces a triplet of random digits. Photographs of these dice are on the cover of Birger Jansson's valuable monograph *Random Number Generators*, written in English and published in Sweden in 1966.

The earliest-known cubical dice, found in Egyptian tombs predating 2000 B.C., are not uniform in size, material, or the way the sides are numbered, although many are identical with modern dice (on which the digits 1 through 6 are placed so that opposite sides add to 7). If that arrangement is not required, there are 30 ways a cube's face can be spotted to represent 1 through 6, counting mirror-reflection forms as different (but not taking into consideration the two different orientations of the 2, 3, and 6 spots, which, in their traditional patterns, lack fourfold symmetry). If opposite sides total 7, as they do on all modern dice, there are just two ways of arranging the numbers, each a mirror reflection of the other.

All Western dice are now made with the same handedness. If you hold a die so that you see its 1–2–3 faces, the numbers go counterclockwise. Dice of both handedness are sold today in Japan [*see Figure 126*]. The type matching Western dice is

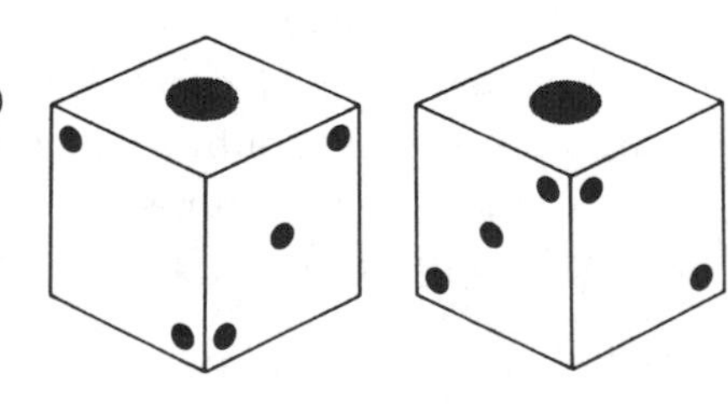

FIGURE 126
Western-style Japanese die (left) and die for Mah-Jongg (right)

used in all Japanese games except Mah-Jongg, which uses dice that—as Lewis Carroll's Alice put it—"go the other way." Dice of either handedness are sold in two forms: the Western, in which all the spots are black, and the traditional Japanese form, in which the ace is very large, deeply indented, and colored red. Chinese and Korean dice also have the large red spot, and in addition the spots on the 4 side are also red. Red sides seem to play no role in Chinese and Korean games except in determining who throws first, the first player being the one who rolls the most red spots. The origin of the red coloring is not known. Stewart Culin, in his privately printed monograph *Chinese Games with Dice* (Philadelphia, 1889), recounted some old Chinese myths that explain the red coloring, but he himself believed it derives from earlier Indian dice.

A professional dice gambler is often so well acquainted with the handedness of modern dice—that is, with exactly how any triplet of numbers is arranged around a corner—that if you show him a die with your finger and thumb covering any two opposite sides, he can immediately tell you which number is under the finger and which under the thumb. This knowledge is useful in detecting a common variety of crooked dice known in the trade as "tops." Tops are cubes mis-spotted so that each bears only three numbers; pairs of identical numbers are on opposite sides. Since no more than three sides of a cube can be seen at one time, a pair of tops resting on a surface appear perfectly normal to all players. There is no way, however, to make a die of this sort so that every triplet of faces goes the "right way." If the reader will take a sugar cube and pencil on it any three pairs of digits that surround a corner on a standard die, putting duplicate numbers on opposite faces, and then inspect the die by viewing each of its eight corners, he will find that at four corners the three faces go the "wrong way" when compared with a modern die. This means that the probability of such a die's falling with its three visible faces showing the wrong handedness is exactly 1/2. When this occurs, a knowledgeable gambler immediately recognizes the die as mis-spotted.

Tops are made in a variety of combinations so that the bust-out man—a man who is expert at secretly switching tops in and out of games—can "bust into" the game whatever kind of tops is needed at the moment. (The essential rules of craps are: If the shooter rolls a "natural" [7 or 11] on his first roll, he wins at once. If he throws a "crap" [2, 3, or 12], he loses. Any other number becomes his "point," and he continues to throw until he either wins by making his point or loses by throwing a 7 before making his point. All of this is accompanied by much betting of various kinds between players, the nature of which depends on whether the game is informal or played in a casino.) For example, if the shooter is trying for a point of 4, 6, 8, or 10, a 1–3–5 and 2–4–6 pair of tops will not form any of those numbers, and so he is sure to "seven out" before he makes his point. Tops of this kind are called "misses." Tops designed to make points and incapable of throwing a 7 (for example, 1–3–5 and 1–3–5) are known as "hits."

Tops cannot be left long in a game because the danger of detection is too great, and so a bust-out man has to work fast, as well as continually and indetectably. "Good bust-out men die young," writes gambling authority John Scarne in *Scarne's Complete Guide to Gambling* (1961); "it's hard on the nervous system."

A "one-way top" is a die with only one number (usually 2 or 5) duplicated. Such dice can be used in pairs or in combination with a legitimate die to give milder percentages, but they are harder to detect and sometimes are left in a game for hours. "Door pops" are still advertised in the catalogues of crooked-gambling-supply firms but are strictly for sucker customers. No professional cheat would consider using them. One set always craps out because one cube bears only aces and deuces, the other only aces. Another set, with 6's and 2's on one die and all 5's on the other, always sevens or elevens. As Scarne puts it, they can be used only on "soft marks" (extremely gullible suckers) "for night play under an overhead light when the chumps can't see anything but the top surfaces of the dice. Strictly for use by cheats who don't know what a set of real tops is."

Mis-spotting is only one of dozens of ways dice can be

"gaffed" for cheating. They can be loaded in many ingenious ways: they can be shaped somewhat like bricks; certain sides can be made a trifle convex, causing the cube to favor the flat sides, or sides can be made slightly concave to create suction on smooth, hard surfaces. Edges can be beveled to alter percentages. "Capped dice" are made with certain faces bouncier than others. "Slick dice" have some faces smoothed and others roughened. Magnetic dice are loaded to roll normally unless an electromagnet concealed under the rolling surface is turned on. Ordinary dice are best tested for loads by dropping them in water many times to see if certain faces show more often than they should. The interested reader will find all this and much more explained in fascinating detail in the standard work on cheating at dice, *Scarne on Dice*, by John Scarne and Clayton Rawson (ninth edition, 1968).

Dicing was enormously popular in ancient Greece and Rome, particularly among the upper classes, and during the Middle Ages it was a favorite time-waster for both knights and the clergy. There were even medieval dicing schools and guilds. In the United States today the most popular dice game is craps. Apparently it dates from the early 1890's, when blacks in the New Orleans area simplified the complicated rules of the English game of hazard. (Dice are still facetiously called "African dominoes.") Then, like jazz, craps spread up the Mississippi River and fanned out over the continent. The big gambling casinos did not take it up until near the end of the nineteenth century. Today it gets faster action than any other casino game. Many players believe the shooter has a 50–50 chance of winning, but it is not hard to prove that the odds are slightly against him. To be precise, the shooter's winning probability is exactly 244/495, or .493+.

It is easy to go wrong in figuring the probability of dice throws. In the tenth chapter of the last book of Rabelais's *Gargantua and Pantagruel* as translated by Jacques Le Clercq (Modern Library, 1944), the adventurers visit Sharper's Island, formed of two enormous cubic blocks of dazzling white bone. "Our navigator informed us," says Pantagruel, "that these cube-shaped white rocks had caused more shipwrecks, entailing a greater loss of life and property, than . . . Scylla and Charybdis." Dice were often called "the devil's bones," and Rabelais has Sharper's Island inhabited by 20 devils of chance, one for each combination of two dice, from Double Six, the largest devil, to Double Aces, the smallest. Actually there are 21 such combinations, a figure given correctly in other translations. The chart in Figure 127 shows the 6 × 6, or 36, different ways two dice can fall. Inspection reveals 21 different combinations. With this basic chart one can quickly calculate the probability of throwing any sum from 2 through 12. Note that 7 can be made in six ways, more than any other sum. The probability of throwing a 7 is therefore 6/36, or 1/6. It is the easiest of all sums to make.

William Saroyan, in a fine short story about crap shooting called "Two Days Wasted in Kansas City," speaks of 4, the point he is trying to make, as "one of the toughest numbers in the world." The chart proves he is right. Two and 12 are the hardest sums to roll, since each can be made in only one way (probability 1/36), but neither 2 nor 12 can be a point. Three and 11 come next, with probabilities of 2/36, or 1/18, each, but 3 is a crap and 11 a natural and so neither of them can be a point either. The hardest points to make are "Little Joe" (4) and "Big Dick" (10). Since each can be made in three ways, the probability of throwing each is 3/36, or 1/12.

Some of the greatest mathematicians have gone astray in calculating dice odds. Leibniz thought 11 and 12 had equal probabilities because each could be made with only one combination of two dice; he failed to consider that 12 can be made in only one way, whereas 11 can be made with either die 6 and the other 5, making 11 twice as easy to throw as 12. The Greeks and Romans preferred games with three dice, and Plato, in his *Laws* (Book 12), cited 3 and 18 as the most difficult sums to roll with three dice. They are the only sums that can be made in only one way (1–1–1 and 6–6–6). Since there are 6 × 6 × 6, or 216, equally probable ways of rolling three dice, the probability of making a 3 is 1/216. The same holds for 18. That 3 and 18

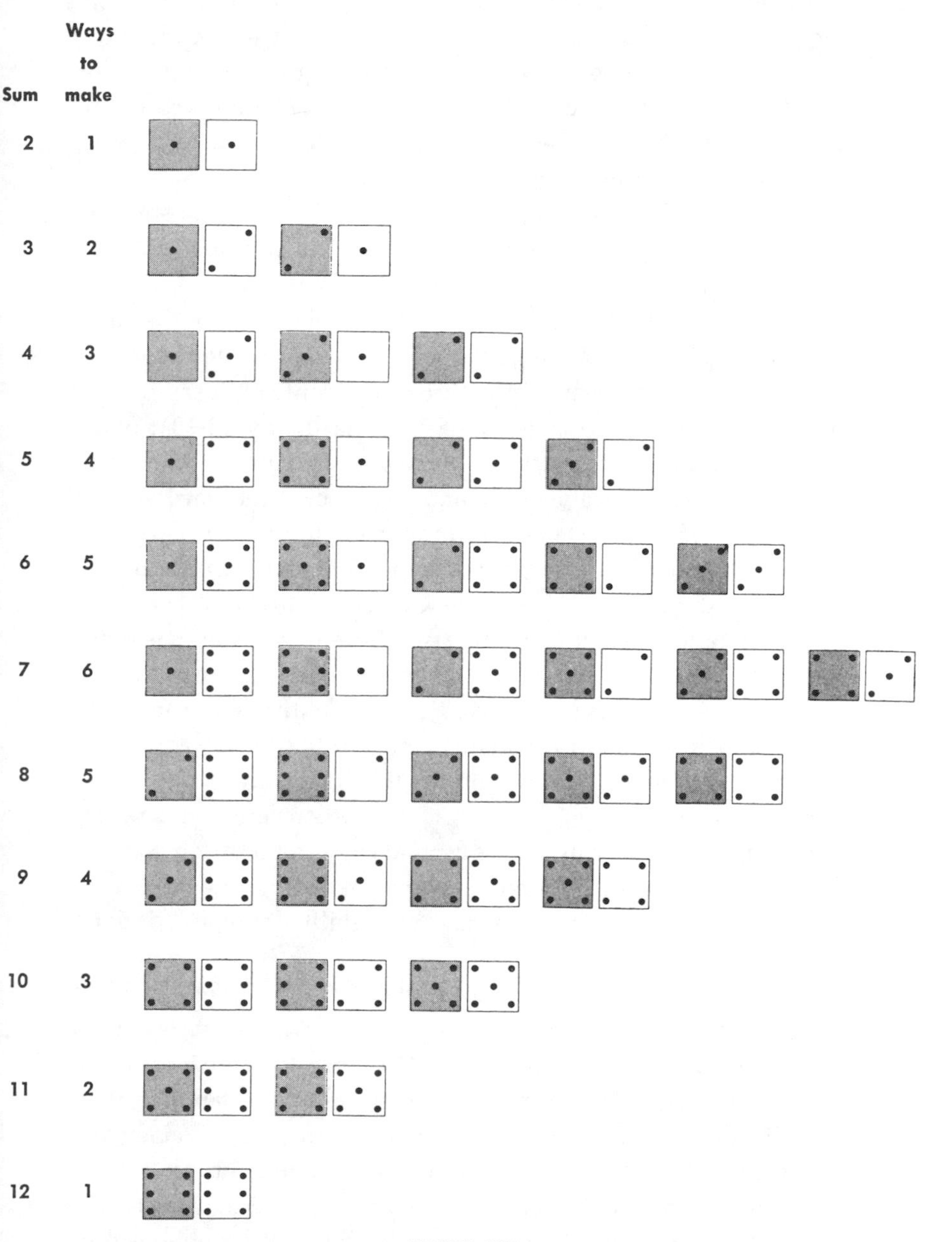

FIGURE 127
The 36 equally probable ways two dice (shaded and white) can fall

were the two most difficult throws was well known to both the Greeks and the Romans. The Greeks called 6–6–6 "Aphrodite" and 1–1–1 "the dog," terms corresponding to our slang of "snake eyes" for 1–1 and "boxcars" for 6–6. There are many references to these and other dicing terms in Greek and Latin literature. The Roman emperor Claudius even wrote a book called *How to Win at Dice,* but unfortunately it did not survive.

Sucker bets are bets professional gamblers like to spring on marks because they can be made at odds that seem to favor the mark but actually do not. In craps, for instance, one might guess that it would be just as easy, if the point is 4, for the shooter to make it the "hard way" (by throwing identical faces on the dice, in this case 2–2) as to make 6 the hard way (3–3). Now, it is true that the probability of throwing any sum the hard way is 1/36, but the probability of making a point the hard way is altogether different. There are three ways to make 4. Only one (2–2) is the hard way. The shooter fails to make his point the hard way if he throws 3–1 or 1–3, or if he throws 7 before he makes 4. Since he can roll 7 in six different ways he has eight ways to lose and one to win, and so the odds are eight to one against making 4 the hard way. Put differently, his probability of doing it is 1/9. Now consider making 6 the hard way. There are five ways to roll 6. Only one is the hard way. The shooter can lose by throwing 6 in any of the other four ways or by throwing 7 in any of its six ways, making a total of ten ways to lose and one to win. The odds against making 6 the hard way are therefore ten to one and the probability of doing it goes down to 1/11.

One of the oldest and subtlest sucker bets goes like this: The hustler first bets even money that the mark will throw an 8 before he throws a 7. The mark, knowing that 7 is easier to make than 8, quickly accepts such bets, which he tends to win. The hustler then switches from 8 to 6, betting even money the mark will throw 6 before 7. Again the hustler tends to lose because 6, like 8, can be made in only five ways as against six ways for 7. Now comes the big swindle. The hustler, who is pretending to be an ignoramus about dice odds, decides to bet even money again, at much higher stakes, that the mark will make *both* 8 and 6 before he throws *two* 7's. This seems to be just as good a bet to the sucker as before; actually the odds now make a surprising shift to favor the hustler. If he had specified the order of the two numbers, first a 6 or first an 8, the odds would have been against him as before. But because the shooter can roll either sum first, it turns out that the hustler has a probability of 4,225/7,744—a bit better than 1/2—of winning.

Here are three easy dice puzzles:

1. A magician turns his back and asks someone to roll three standard dice and add the top faces. The spectator then picks up any die and adds its bottom number to that total. The same die is rolled again and its top number is added to the previous total. The magician turns around for the first time to glance at the dice. Although he has no way of knowing which cube was picked for the extra roll, he is able to state the final total correctly. How does he do it?

2. The same mis-spotted die is shown in three views in Figure 128. How many spots are opposite the 6? (The problem is from *Puzzles in Math and Logic*, by Aaron J. Friedland [Dover, 1970].)

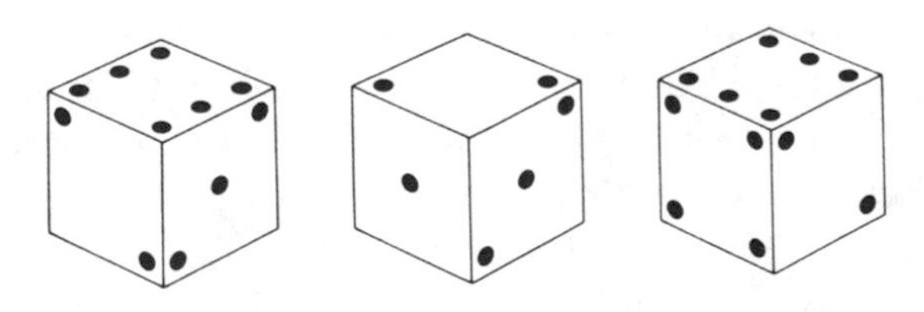

FIGURE 128

How many spots are opposite the 6 on this mis-spotted die?

3. How can two cubes be labeled, each side bearing a number from 1 through 6 or left blank, to make a pair of dice that will throw with equal probability each sum from 1 through 12?

The use of the die as a randomizer has made it a popular literary symbol for chance. We are all familiar with such expressions as "The die is cast" (said to have been uttered by Julius

Caesar after he had made his decision to cross the Rubicon), and the ancient Greeks had a proverb, "The dice of the gods are always loaded." The central dogma of quantum mechanics is that pure chance underlies events on the quantum level; in Einstein's well-known metaphor, quantum mechanics implies that God dices with the universe. It is sometimes argued that even though this may be true on the quantum level, on the macrolevel of human history strictly deterministic laws must still hold. A simple thought experiment provides a dramatic counterexample. Imagine an artificial satellite carrying a hydrogen bomb. The bomb's release is triggered by a Geiger counter click recording the emission of an electron in radioactive decay. If the timing of such a click is pure chance, as quantum theory demands, then pure chance decides which portion of the earth is demolished. Thus we could, in actual practice, make an instant leap from pure chance in the quantum microworld to a major alteration of macroworld history, a thought most disturbing to philosophical determinists.

The view that God dices with human history has found a grimly amusing literary expression in Robert Coover's novel *The Universal Baseball Association, Inc., J. Henry Waugh, Prop.* J. Henry Waugh, whose name suggests Jehovah, is a lonely accountant who lives over a delicatessen. To amuse himself he invents a way of playing imaginary baseball by rolling three dice, assigning certain events to each of the 56 combinations and various sequences of combinations. (Originally he based his games on the 216 ways three dice can fall, using dice of three different colors, but after almost going blind trying to sort out the colors on every throw he shifted to three white dice, considering only their combinations.) Over the months Waugh begins to imagine actual personalities playing on his team until, like so many great characters of fiction—Don Quixote, D'Artagnan, Sherlock Holmes—the players inside Waugh's skull take on a life of their own to the point where they become, in a sense, more real and permanent than Waugh himself. They even begin to wonder if Waugh exists.

One thinks of Pirandello's play *Six Characters in Search of an Author* and Unamuno's earlier novel *Niebla* (*Mist*), in which the protagonist visits the author to protest the author's decision to have him die at the close of the novel, and to remind Unamuno that he too may be only a misty, impermanent dream in the mind of some inconceivably vast roller of the devil's bones.

ANSWERS

THE MAGICIAN names the final total by simply adding seven to the sum of the three top faces of the dice. This total is the sum of the three top faces plus the previous top and bottom of one die. Since opposite sides of a die total seven, the working is obvious.

The trick is a simplification of a trick given by Claude Gaspar Bachet in a 1612 book on mathematical recreations. In Bachet's version someone rolls three dice, adds the faces, picks any two dice, adds their bottom faces to the total, throws the two again, adds their top faces, selects one of the two, adds its bottom face, throws it, adds its top face. In this case the final total is the sum on the top faces plus 21.

The answer to the second problem is that the mis-spotted die shown from three angles must have a deuce opposite the six. Figure 129 shows how the die would look if its faces were unfolded.

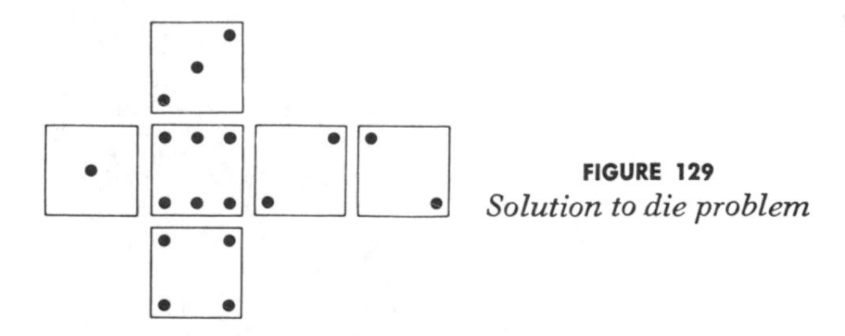

FIGURE 129
Solution to die problem

The third problem is from *100 Brain-twisters*, by D. St. P. Barnard, a collection of original puzzles, published in England in 1966 and distributed here by D. Van Nostrand. Since there are 36 ways two dice can fall, if sums from 1 through 12 are to be made with equal probability, each must be made in three ways. The only way to make 12 in three ways is by having a 6 on one die and three 6's on the other. The only way to make 1 in three ways is to have a 1 on one die and three blanks on the other. This leads to the only solution: one standard die, the other with three 6's and three blanks.

The method applies to any of the five regular solids. For example, consider a pair of icosahedral dice. If the numbers 1 through 20 appear on one icosahedron and the other die has 10 blank faces and 10 faces bearing 20's, the pair will throw with equal probability any sum from 1 through 40.

CHAPTER 19

Everything

A curious thing about the ontological problem is its simplicity. It can be put in three Anglo-Saxon monosyllables: "What is there?" It can be answered, moreover, in a word—"Everything."

—Willard Van Orman Quine, "On What There Is"

The topic of the first chapter of this book is "Nothing." I have nothing more to say about nothing, or about "something," since everything I know about something was said when I wrote about nothing. But "everything" is something altogether different.

Let us begin by noting the curious fact that some things, namely ourselves, are such complicated patterns of waves and particles that they are capable of wondering about everything. "What is man in nature?" asked Pascal. "A nothing in comparison with the infinite, an all in comparison with the nothing, a mean between nothing and everything."

In logic and set theory "things" are conveniently diagrammed with Venn circles. In Figure 130 the points inside circle a represent humans. The points inside circle b stand for feathered animals. The overlap, or intersection set, has been darkened to show that it has no members. It is none other than our old friend the empty set.

So far, so clear. What about the points on the plane outside

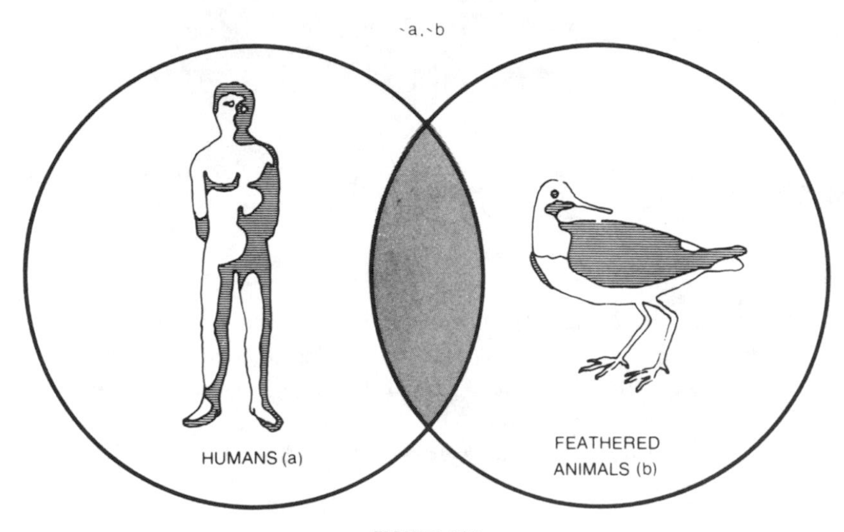

FIGURE 130
A Venn diagram for "No humans have feathers"

the two circles? Obviously they represent things that are not *a* and not *b*, not human and not feathered, but how far-ranging is this set? To clarify the question Augustus De Morgan invented the phrase "universe of discourse." It is the range of all the variables with which we are concerned. Sometimes it is explicitly defined, sometimes tacitly assumed, sometimes left fuzzy. In set theory it is made precise by defining what is called the universal set, or, for short, the universe. This is the set with a range that coincides with the universe of discourse. And that range can be whatever we want it to be.

With the Venn circles *a* and *b* we are perhaps concerned only with living things on the earth. If this is so, that is our universe. Suppose, however, we expand the universe by adding a third set, the set of all typewriters, and changing *b* to all feathered objects. As Figure 131 shows, all three intersection sets are empty. It is the same empty set, but the range of the null set has also been expanded. There is only one "nothing," but a hole in the ground is not the same as a hole in a piece of

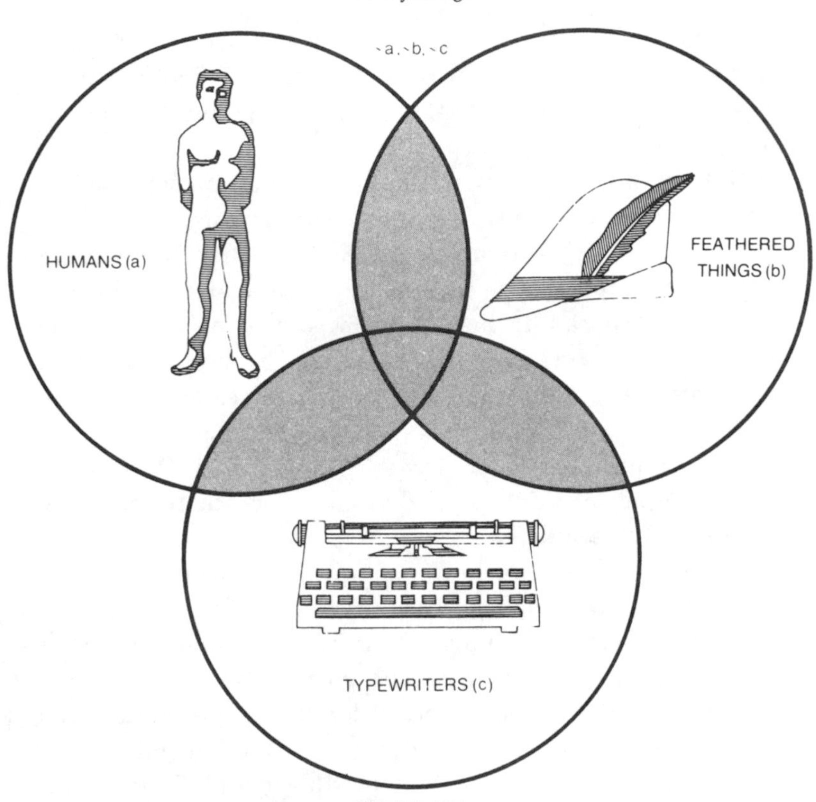

FIGURE 131
A Venn diagram for three sets

cheese. The complement of a set k is the set of all elements in the universal set that are not in k. It follows that the universe and the empty set are complements of each other.

How far can we extend the universal set without losing our ability to reason about it? It depends on our concern. If we expand the universe of Figure 130 to include all concepts, the intersection set is no longer empty because it is easy to imagine a person growing feathers. The proofs of Euclid are valid only if the universe of discourse is confined to points in a Euclidean plane or in 3-space. If we reason that a dozen eggs can be equally divided only between one, two, three, four, six, or twelve people, we are reasoning about a universal set that ranges over the integers. John Venn (who invented the Venn diagram) likened the universe of discourse to our field of vision. It is what we are looking at. We ignore everything behind our head.

Nevertheless, we can extend the universe of discourse amazingly far. We certainly can include abstractions such as the number 2, pi, complex numbers, perfect geometric figures, even things we cannot visualize such as hypercubes and non-Euclidean spaces. We can include universals such as redness and cowness. We can include things from the past or in the future and things real or imaginary, and can still reason effectively about them. Every dinosaur had a mother. If it rains next week in Chicago, the old Water Tower will get wet. If Sherlock Holmes had actually fallen off that cliff at Reichenbach Falls, he would have been killed.

Suppose we extend our universe to include every entity that can be defined without logical contradiction. Every statement we can make about that universe, if it is not contradictory, is (in a sense) true. The contradictory objects and statements are not allowed to "exist" or be "true" for the simple reason that contradiction introduces meaninglessness. When a philosopher such as Leibniz talks about "all possible worlds," he means worlds that can be talked about. You can talk about a world in which humans and typewriters have feathers. You cannot say anything sensible about a square triangle or an odd integer that is a multiple of 2.

Is it possible to expand our universe of discourse to the ultimate and call it the set of all possible sets? No, this is a step we cannot take without contradiction. Georg Cantor proved that the cardinal number of any set (the number of its elements) is always lower than the cardinal number of the set of all its subsets. This is obvious for any finite set (if it has n elements, it must have 2^n subsets), but Cantor was able to show that it also applies to infinite sets. When we try to apply this theorem to everything, however, we get into deep trouble. The set of all sets must have the highest aleph (infinite number) for its cardinality; otherwise it would not be everything. On the other hand, it cannot have the highest aleph because the cardinality of its subsets is higher.

When Bertrand Russell first came across Cantor's proof that there is no highest aleph, and hence no "set of all sets," he did not believe it. He wrote in 1901 that Cantor had been "guilty of a very subtle fallacy, which I hope to explain in some future work," and that it was "obvious" there had to be a greatest aleph because "if everything has been taken, there is nothing left to add." When this essay was reprinted in *Mysticism and Logic* sixteen years later, Russell added a footnote apologizing for his mistake. ("Obvious" is obviously a dangerous word to use in writing about everything.) It was Russell's meditation on his error that led him to discover his famous paradox about the set of all sets that are not members of themselves.

To sum up, when the mathematician tries to make the final jump from lots of things to everything, he finds he cannot make it. "Everything" is self-contradictory and therefore does not exist!

The fact that the set of all sets cannot be defined in standard (Zermelo-Fraenkel) set theory, however, does not inhibit philosophers and theologians from talking about everything, although their synonyms for it vary: being, *ens*, what is, existence, the absolute, God, reality, the Tao, Brahman, *dharmakaya*, and so on. It must, of course, include everything that was, is and will be, everything that can be imagined and everything totally beyond human comprehension. Nothing is also part of everything. When the universe gets this broad, it is difficult to think of anything meaningful (not contradictory) that does not in some sense exist. The logician Raymond Smullyan, in one of his several hundred marvelous unpublished essays, retells an incident he found in Oscar Mandel's book *Chi Po and the Sorcerer: A Chinese Tale for Children and Philosophers*. The sorcerer Bu Fu is giving a painting lesson to Chi Po. "No, no!" says Bu Fu. "You have merely painted what *is*. Anybody can paint what is! The real secret is to paint what isn't!" Chi Po, puzzled, replies: "But what is there that isn't?"

This is a good place to come down from the heights and consider a smaller, tidier universe, the universe of contemporary cosmology. Modern cosmology started with Einstein's model of a closed but unbounded universe. If there is sufficient mass in the cosmos, our 3-space curves back on itself like the surface of a sphere. (Indeed, it becomes the 3-space hypersurface of a 4-space hypersphere.) We now know that the universe is expanding from a primordial fireball, but there does not seem to be enough mass for it to be closed. The steady-state theory generated much discussion and stimulated much valuable scientific work, but it now seems to have been eliminated as a viable theory by such discoveries as that of the universal background radiation (which has no reasonable explanation except that it is radiation left over from the primordial fireball, or "big bang").

The large unanswered question is whether there is enough mass hidden somewhere in the cosmos (in black holes?) to halt the expansion and start the universe shrinking. If that is destined to happen, the contraction will become runaway collapse, and theorists see no way to prevent the universe from entering the "singularity" at the core of a black hole, that dreadful spot where matter is crushed out of existence and no known laws of physics apply. (For a superb painting of a black hole, see Figure 2.) Will the universe disappear like the fabled Poof Bird, which flies backward in ever decreasing circles until—poof!—it vanishes into its own anus? Will everything go through the black hole to emerge from a white hole in some completely different spacetime? Or will it manage to avoid the singularity and give rise to another fireball? If reprocessing is possible, we have a model of an oscillating universe that periodically explodes, expands, contracts, and explodes again.

Among physicists who have been building models of the universe John Archibald Wheeler of Princeton University has gone further than anyone in the direction of everything. In Wheeler's wild vision our universe is one of an infinity of universes that can be regarded as embedded in a strange kind of space called superspace.

In order to understand (dimly) what Wheeler means by

superspace let us start with a simplified universe consisting of a line segment occupied by two particles, one black and one gray [*see Figure 132, top*]. The line is one-dimensional, but the particles move back and forth (we allow them to pass through each other) to create a spacetime of two dimensions: one of space and one of time.

There are many ways to graph the life histories of the two particles. One way is to represent them as wavy lines, called world lines in relativity theory, on a two-dimensional spacetime graph [*see Figure 132, bottom*]. Where was the black particle at time k? Find k on the time axis, move horizontally to the black particle's world line, then move down to read off the particle's position on the space axis.

To see how beautifully the two world lines record the history of our infant universe, cut a slot in a file card. The slot should be as long as the line segment and as wide as a particle. Place the card at the bottom of the graph where you can see the universe through it. Move the card upward slowly. Through the slot you will see a motion picture of the two particles. They are born at the center of their space, dance back and forth until they have expanded to the limits, and then dance back to the center, where they disappear into a black hole.

In kinematics it is sometimes useful to graph the changes of a system of particles as the motion of a single point in a higher space called configuration space. Let us see how to do this with our two particles. Our configuration space again is two-dimensional, but now both coordinates are spatial. One coordinate is assigned to the black particle and the other to the gray particle [*see Figure 133*]. The positions of both particles can be represented by a single point called the configuration point. As the point moves, its coordinate values change on both axes. One axis locates one particle, the other axis the other particle. The trajectory traced by the moving point corresponds to the changing pattern of the system of particles; conversely, the history of the system determines a unique trajectory. It is not a spacetime graph. (Time enters later as an added parameter.) The line cannot form branches because that would split each particle in

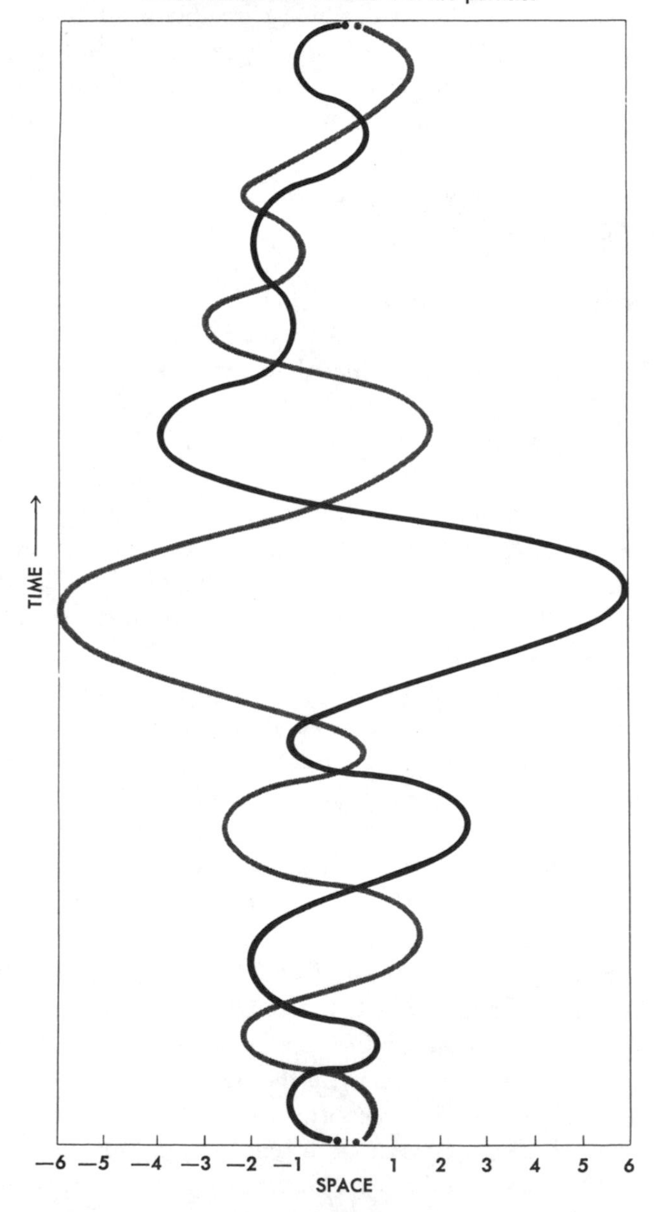

FIGURE 132

A space-time graph of a two-particle cosmos from birth to death

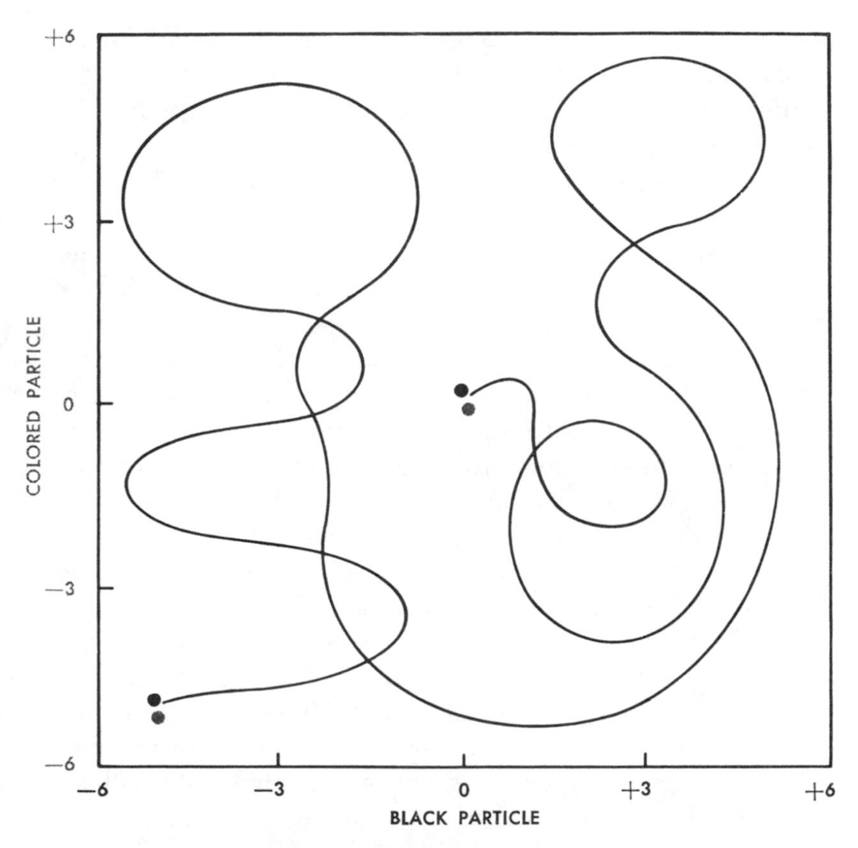

FIGURE 133
A configuration-space graph of the history of two particles in a one-dimensional universe

two. It may, however, intersect itself. If a system is periodic, the line will be a closed curve. To transform the graph into a spacetime graph we can, if we like, add a time coordinate and allow the point to trace a curve in three dimensions.

The technique generalizes to a system of N particles in a space with any number of dimensions. Suppose we have 100 particles in our little line-segment cosmos. Each particle has one degree of freedom, so our configuration point must move in a space of 100 dimensions. If our universe is a system of N particles on a plane, each particle has two degrees of freedom, so our configuration space must be a hyperspace of $2N$ dimensions. In 3-space a particle has three degrees of freedom, so the configuration space must have $3N$ dimensions. In general the hyperspace has an order equal to the total degrees of freedom in the system. Add another coordinate for time and the space becomes a spacetime graph.

Unfortunately the position of a configuration point at any instant does not enable us to reconstruct the system's past or predict its future. Josiah Willard Gibbs, working on the thermodynamics of molecules, found a slightly more complicated space in which he could graph a system of molecules so that the record was completely deterministic. This is done by assigning six coordinates to each molecule: three to determine position and three to specify momentums. The movement of a single phase point in what Gibbs called a "phase space" of $6N$ dimensions will record the life history of N particles. Now, however, the position of the phase point provides enough information to reconstruct (in principle) the entire previous history of the system and to predict its future. As before, the trajectory cannot branch, but now it also cannot intersect itself. An intersection would mean that a state could be reached from two different states, and could lead to two different states, but both possibilities are ruled out by the assumption that position and momentums (which include a vector direction) fully determine the next state. The curve may still loop, however, indicating that the system is periodic.

Our universe, with its non-Euclidean spacetime and its quantum uncertainties, cannot be graphed in anything as simple as phase space, but Wheeler has found a way to do it in superspace. Like configuration space, superspace is timeless, but it has an infinity of dimensions. A single point in superspace has an infinite set of coordinates that specify completely the structure of our non-Euclidean 3-space: its size, the location of every particle, and the structure of every field (including the curvature of space itself) at every point. As the superpoint moves, its changing coordinate numbers describe how our universe changes, not failing to take into account the role of observers' frames of reference in relativity and the probability parameters of quantum mechanics. The motion of the superpoint gives the entire history of our universe.

At the same time (whatever that means!) that the present drama of our cosmos is being acted on the stage of superspace countless other superpoints, representing other 3-space universes, are going through their cycles. Superpoints close to one another describe universes that most resemble one another, like the parallel worlds that H. G. Wells introduced into science fiction with his *Men Like Gods*. These parallel universes, cut off from one another because they occupy different slices of superspace, are continually bursting into spacetime through a singularity, flourishing for a moment of eternity, then vanishing back through a singularity into the pure and timeless "pregeometry" from whence they came.

Whenever such a cosmos explodes into being, random factors generate a specific combination of logically consistent (Leibniz called them compossible) particles, constants, and laws. The resulting structure has to be tuned exceedingly fine to allow life. Alter the fine-structure constant a trifle either way and a sun such as ours becomes impossible. Why are we here? Because random factors generated a cosmic structure that allowed us to evolve. An infinity of other universes, not so finely tuned, are living and dying without there being anyone in them capable of observing them.

These "meaningless" universes, meaningless because they contain no participator-observers, do not even "exist" except in the weak sense of being logically possible. Bishop Berkeley said that to exist is to be perceived, and Charles Sanders Peirce maintained that existence is a matter of degree. Taking cues from both philosophers, Wheeler argues that only when a universe develops a kind of self-reference, with the universe and its observers reinforcing one another, does it exist in a strong sense. "All the choir of heaven and furniture of earth have no substance without a mind" was how Berkeley put it.

As far as I can tell, Wheeler does not take Berkeley's final step: the grounding of material reality in God's perception. Indeed, the fact that a tree seems to exist in a strong sense, even when no one is looking at it, is the key to Berkeley's way of proving God's existence. Imagine a god experimenting with billions of cosmic models until he finds one that permits life. Would not these universes be "out there," observed by the deity? There would be no need for flimsy creatures like ourselves, observing and participating, to confer existence on these models.

Wheeler seems anxious to avoid this view. He argues that quantum mechanics requires participator-observers in the universe regardless of whether there is an outside observer or not. In one of his metaphors, a universe without internal observers is like a motor without electricity. The cosmos "runs" only when it is "guaranteed to produce somewhere, and for some little length of time in its history-to-be, life, consciousness and observership." Internal observers and the universe are both essential to the existence of each other, even if the observers exist only in a potential sense. This raises unusual questions. How strongly does a universe exist before the first forms of life evolve? Does it exist in full strength from the moment of big bang, or does its existence get stronger as life gets more complex? And how strong is the existence of a galaxy, far removed from the Milky Way, in which there may be no participator-observers? Does it exist only when it is observed by life in another galaxy? Or is the universe so interconnected that the observation of a minute portion of it supports the existence of all the rest?

There is a famous passage in which William James imagines a thousand beans flung onto a table. They fall randomly, but our eyes trace geometrical figures in the chaos. Existence, wrote James, may be no more than the order which our consciousness singles out of a disordered sea of random possibilities. This seems close to Wheeler's vision. Reality is not something out there, but a process in which our consciousness is an essential part. We are not what we are because the world is what it is, but the other way around. The world is what it is because we are what we are.

When relativity theory first won the day, many scientists and philosophers with a religious turn of mind argued that the new theory supported such a view. The phenomena of nature, said James Jeans, are "determined by us and our experience rather than by a mechanical universe outside us and independent of us." The physical world, wrote Arthur Stanley Eddington, "is entirely abstract and without 'actuality' apart from its linkage to consciousness." Most physicists today would deny that relativity supports this brand of idealism. Einstein himself vigorously opposed it. The fact that measurements of length, time, and mass depend on the observer's frame of reference in no way dilutes the actuality of a spacetime structure independent of all observers.

Nor is it diluted by quantum mechanics. What bearing does the statistical nature of quantum laws have on the independent existence of a structure to which those laws apply whenever it is observed? The fact that observations alter state functions of a system of particles does not entail that there is nothing "out there" to be altered. Einstein may have thought that quantum mechanics implies this curious reduction of physics to psychology, but there are not many quantum experts today who agree.

In any case, belief in an external world, independent of human existence but partly knowable by us, is certainly the simplest view and the one held today by the vast majority of scientists and philosophers. As I have suggested, to deny this common-sense attitude adds nothing of value to a theistic or pantheistic faith. Why adopt an eccentric terminology if there is no need for it?

But this is not the place for debating these age-old questions. Let me turn to a strange little book called *Eureka: A Prose Poem*, written by Edgar Allan Poe shortly before his death. Poe was convinced that it was his masterpiece. "What I have propounded will (in good time) revolutionize the world of Physical & Metaphysical Science," he wrote to a friend. "I say this calmly—but I say it." In another letter he wrote, "It is no use to reason with me *now;* I must die. I have no desire to live since I have done *Eureka.* I could accomplish nothing more."

(I quote from excellent notes in *The Science Fiction of Edgar Allan Poe,* edited by Harold Beaver, Penguin Books, 1976.)

Poe wanted his publisher, George P. Putnam, to print 50,000 copies. Putnam advanced Poe fourteen dollars for his "pamphlet," and printed 500 copies. Reviews were mostly unfavorable. To this day the book seems to have been taken seriously only in France, where it had been translated by Baudelaire. Now suddenly, in the light of current cosmological speculation, Poe's prose poem is seen to contain a vast vision that is essentially a theist's version of Wheeler's cosmology! As Beaver points out, the "I" in Poe's "Dreamland" has become the universe itself:

By a route obscure and lonely,
Haunted by ill angels only,
Where an Eidolon, named NIGHT,
On a black throne reigns upright,
I have reached these lands but newly
From an ultimate dim Thule—
From a wild weird clime that lieth, sublime,
Out of SPACE—*out of* TIME.

A universe begins, said Poe, when God creates a "primordial particle" out of nothing. From it matter is "irradiated" spherically in all directions, in the form of an "inexpressibly great yet limited number of unimaginably yet not infinitely minute atoms." As the universe expands, gravity slowly gains the upper hand and the matter condenses to form stars and planets. Eventually gravity halts the expansion and the universe begins to contract until it returns again to nothingness. The final "globe of globes will instantaneously disappear" (how Poe would have exulted in today's black holes!) and the God of our universe will remain "all in all."

In Poe's vision each universe is being observed by its own deity, the way your eye watched the two particles dance in our created world of 1-space. But there are other deities whose eyes watch other universes. These universes are "unspeakably distant" from one another. No communication between them is possible. Each of them, said Poe, has "a new and perhaps totally different series of conditions." By introducing gods Poe implies that these conditions are not randomly selected. The fine-structure constant is what it is in our universe because our deity wanted it that way. In Poe's superspace the cyclical birth and death of an infinity of universes is a process that goes on "for ever, and for ever, and for ever; a novel Universe swelling into existence, and then subsiding into nothingness at every throb of the Heart Divine."

Did Poe mean by "Heart Divine" the God of our universe or a higher deity whose eye watches all the lesser gods from some abode in supersuperspace? Behind Brahma the creator, goes Hindu mythology, is Brahman the inscrutable, so transcendent that all we can say about Brahman is *Neti neti* (not that, not that). And is Brahman being observed by a supersupersuper-eye? And can we posit a final order of superspace, with its Ultimate Eye, or is that ruled out by the contradiction in standard set theory of the concept of a greatest aleph?

This is the great question asked in the final stanza of the Hymn of Creation in the *Rig Veda*. The "He" of the stanza is the impersonal One who is above all gods:

Whether the world was made or was self-made,
He knows with full assurance, He alone,
Who in the highest heaven guards and watches;
He knows indeed, but then, perhaps, He knows not!

It is here that we seem to touch—or perhaps we are still infinitely far from touching—the hem of Everything. Let C. S. Lewis (I quote from Chapter 2 of his *Studies in Words*) make the final comment: " 'Everything' is a subject on which there is not much to be said."

Postscript

1 AND 2. NOTHING, AND MORE ADO ABOUT NOTHING

WHEN THE FIRST EDITION of this book was copy edited, an editor asked me if it was necessary to obtain MOMA's (Museum of Modern Art) permission to reprint Ad Reinhardt's all-black painting (Figure 2). I convinced her it was not necessary. And as I anticipated, somewhere along the production line I was asked for the missing art of Figure 3. This picture of the null graph, by the way, is reproduced (without credit to the artist) as Figure 257 in the *Dictionary of Mathematics*, edited by E. J. Borowski and J. M. Borwein (London: Collins, 1989).

Meditating on the recent flurry of interest in what is called the "anthropic principle" (on this see Chapter 31 in *Gardner's Whys and Wherefores*, University of Chicago Press, 1989), I suddenly realized that I could answer the superultimate question: Why is there something rather than nothing? Because if there wasn't anything we wouldn't be here to ask the question. I think this points up the essential absurdity of the weak anthropic principle. It's not wrong, but it contributes nothing significant to any philosophical or scientific question. The Danish poet Piet Hein, in one of his "grook" verses, says it this way:

> The universe may
> Be as great as they say,
> But it wouldn't be missed
> If it didn't exist.

Lakenan Barnes, an attorney in Missouri, reminded me that Joshua was the son of Nun (Joshua 1:1), that "love" in tennis

FIGURE 134

Henry Moore's "Nuclear Energy," at the University of Chicago, commemorates the site where Enrico Fermi achieved the first sustained nuclear reaction in 1942. By an adroit use of holes, Moore has combined a mushroom cloud, the eye sockets of a skull, the look of an embryo, and the vaulting of a cathedral.

means nothing, that the doughnut's hole is the "dough naught." He also passed along a quatrain of his that had appeared in the St. Louis *Post-Dispatch* (July 7, 1967):

> In the world of math
> That Man has wrought,
> The greatest gain
> Was the thought of naught.

Some readers were mystified by the chapter's epigraph. It is the second sentence of Heath's article on nothing in the *Encyclopedia of Philosophy*. Like Lewis Carroll, in the second Alice book, Heath is taking Nobody to be the name of a person. Here is the sentence in the context of Heath's playful opening paragraph:

> NOTHING is an awe-inspiring yet essentially undigested concept, highly esteemed by writers of a mystical or existentialist tendency, but by most others regarded with anxiety, nausea, or panic. Nobody seems to know how to deal with it (he would, of course), and plain persons generally are reported to have little difficulty in saying, seeing, hearing, and doing nothing. Philosophers, however, have never felt easy on the matter. Ever since Parmenides laid it down that it is impossible to speak of what is not, broke his own rule in the act of stating it, and deduced himself into a world where all that ever happened was nothing, the impression has persisted that the narrow path between sense and nonsense on this subject is a difficult one to tread and that altogether the less said of it the better.

4. FACTORIAL ODDITIES

CAN A FACTORIAL greater than 1! be a square number? Can the sum of the first m factorials (except for $m = 1$ and 3) be a square? (We take the first factorial to be 0!) Simple proofs of "no" for both questions will be found in the note by David Silverman cited in the bibliography.

Douglas Hofstadter wrote to ask the basis for this curious sequence: 0, 1, 2, 720! Answer: 0, 1!, 2!!, 3!!!, ...

Gustavus J. Simmons, in the two notes cited in the bibliography, conjectures that 3!, 4!, 5!, and 6! are the only four factorials

ACROSS

1. "____ to it"
6. Bestseller *The ____ Book*
15. What Lady Godiva wore
16. Opposite of something
17. What Old Mother Hubbard found in her cupboard
18. Good for ____
19. "Here goes ____ !"
20. Zero
21. ____ succeeds like success
23. To say ____ of
24. Not anything
25. Naught
26. Trifling or inane remark
29. For ____ (free)
30. Love, in tennis
31. NIGHT ON (anag.)
32. Nil
33. Come to ____ (fail)
34. Have ____ to do with
35. *Rien*
36. Answer to "What are you doing?"
37. Goose egg
41. Leaving ____ to chance
43. What amnesiacs can remember
44. *Much Ado About* ____
45. No substance
46. Absence of quantity
47. Leave ____ to the imagination
48. "Is ____ sacred?"
49. What's certain besides death and taxes

DOWN

1. What's more fun than GAMES
2. $(12 \div 3) - 2^2$
3. ____ new under the sun
4. Thanks for ____
5. Contents of the null set
6. 0
7. The middle of a doughnut
8. Double or ____
9. ____ to worry about
10. Word with ventured and gained
11. Take ____ for granted
12. Bankrupt's net worth
13. Nihil
14. ". . . and ____ but the truth"
20. Utter insignificance
22. Zilch
25. "Wise men say ____ in dangerous times" (John Selden)
26. What's in a vacuum
27. In ____ flat
28. What moves faster than the speed of light
29. Know-____ Party
31. What, subtracted from itself, leaves itself
33. Better than ____
35. The emperor's "new clothes"
37. "____ doing"
38. "I have ____ to wear" (familiar complaint)
39. Nonentity
40. Leave ____ behind
42. This:
45. What to do to complete this puzzle

FIGURE 135

This crossword puzzle by Will Shortz appeared in the March/April 1979 issue of Games magazine as an April Fool's joke. Because the answer to each definition is nothing, the correct solution is to leave the grid blank.

equal to the product of three consecutive digits. (The triplets are $1 \times 2 \times 3$, $2 \times 3 \times 4$, $4 \times 5 \times 6$, and $8 \times 9 \times 10$.) As far as I know, this conjecture remains undecided.

Charles W. Trigg posed this problem in the note cited. The only odd n whose factorial has n digits is 1. What odd n has a factorial with $2n$ digits? The only answer is 267.

Dean Huffman, for his Christmas card, modified the tree shown in Figure 10 (for 105!) by arranging the 25 terminal zeroes in a 5×5 square to make the tree's trunk.

Joseph Madachy informed me that 450! is known as the Arabian Nights factorial because it has 1,001 digits.

I mentioned earlier that no one knows if factorial-plus-one primes are finite or infinite in number. Seventeen are known: $n = 1, 2, 3, 11, 27, 37, 41, 73, 77, 116, 154, 320, 340, 399, 427, 872$, and 1477. The last prime, $1477! + 1$, has 4,042 digits.

It also is not known whether factorial-minus-one primes are finite or infinite in number. Fifteen have been found: $n = 3, 4, 6, 7, 12, 14, 30, 32, 33, 38, 94, 166, 324, 379$, and 469. The last one, $469! - 1$, has 1,051 digits. For this data on both types of factorial primes I am indebted to Samuel Yates, one of the nation's prime prime watchers.

5. THE COCKTAIL CHERRY AND OTHER PROBLEMS

A BREAKTHROUGH ON LANGFORD's problem was obtained by two Japanese mathematicians at the Miyagi Technical College. Takanois Hayasaka and Sadao Saito used a special-purpose calculator to search for quartets that would form Langford sequences. They found three, each of length $4 \times 24 = 96$ numbers. They proved that the minimum value of n (the number of quartets) is 24. The three sequences are given in their note on "Langford Sequences: A Progress Report," in *Mathematical Gazette*, Vol. 63, December 1979, pages 261–262. That same year they reported the results of a computer search for pentuplets through $n = 24$, and sextuplets through $n = 21$. No solutions were found.

In 1980 at Lewis and Clark College, John Miller completed a computer search on the original Langford problem, with doublets, for solutions when $n = 15$. He reports that he found 39,809,640 chains, excluding reversals. More recently, on Nickerson's variant, he found 227,968 solutions for $n = 12$, and 1,520,280 solutions for $n = 13$.

When I said that Eugene Levine found only one solution for triplets when $n = 9$, I did not know that Miller had made an exhaustive computer search that found three solutions for $n = 9$.

1 9 1 6 1 8 2 5 7 2 6 9 2 5 8 4 7 6 3 5 4 9 3 8 7 4 3
1 9 1 2 1 8 2 4 6 2 7 9 4 5 8 6 3 4 7 5 3 9 6 8 3 5 7
1 8 1 9 1 5 2 6 7 2 8 5 2 9 6 4 7 5 3 8 4 6 3 9 7 4 3

Another exhaustive search by Miller turned up five solutions for triplets when $n = 10$.

1 10 1 6 1 7 9 3 5 8 6 3 10 7 5 3 9 6 8 4 5 7 2 10 4 2 9 8 2 4
1 10 1 2 1 4 2 9 7 2 4 8 10 5 6 4 7 9 3 5 8 6 3 10 7 5 3 9 6 8
4 10 1 7 1 4 1 8 9 3 4 7 10 3 5 6 8 3 9 7 5 2 6 10 2 8 5 2 9 6
8 1 10 1 3 1 9 6 3 8 4 7 3 10 6 4 9 5 8 7 4 6 2 5 10 2 9 7 2 5
1 3 1 10 1 3 4 9 6 3 8 4 5 7 10 6 4 9 5 8 2 7 6 2 5 10 2 9 8 7

A recent paper on the general problem is "Exponential Lower Bounds for the Numbers of Skolem and Extremal Langford Sequences," by Jaromir Abram, in *Ars Combinatoria*, Vol. 22, 1986, pages 187–198.

6. DOUBLE ACROSTICS

A. ROSS ECKLER, in a 1986 article in *Word Ways*, a quarterly journal he edits (see bibliography), reported his discovery of a double acrostic earlier than the 1856 one cited by Henry Dudeney. In an 1852 issue of a British periodical called *The Family Friend*, Eckler found two word puzzles that clearly are double and triple acrostics.

7. PLAYING CARDS

RUDOLF ONDREJKA confirmed my statement that the probability of winning the poker bet with 25 randomly chosen cards is indeed high. He shuffled a deck, dealt himself 1,000 samples of 25 cards, and found he could form the five poker hands 986 times. He estimates the probability of winning at 98 to 99 percent. It would be interesting to program a computer to pin down the exact odds.

8. FINGER ARITHMETIC

JAMES ALBERT LINDON (he liked to be called JAL) was a resident of Addlestone, England, where he and a sister ran what he called a "miserable little gift shop." He died in 1979, at age 65, almost penniless, almost blind, and almost unknown. Although we never met, I sorely miss his letters and the unpublished verse and ingenious word play that came with them. His comic poems could be assembled from his friends and correspondents into a marvelous book, but who would publish it?

9. MÖBIUS BANDS

LORRAINE LARISON, a biology professor at the University of Massachusetts, called attention (see bibliography) to a controversy I had not known about. There is some evidence that J. B. Listing, one of topology's pioneers, studied the Möbius strip several years prior to the 1865 paper in which Möbius first published his study of the surface. She cites as a reference *Invitation to Combinatorial Topology*, by M. Fréchet and K. Fan (Prindle, Weber, and Schmidt, 1967), page 29.

Environment Canada, in 1984, appropriated the Möbius strip for its logo to symbolize the recycling of materials. As its press release put it: "Citizen's groups and recycling industries canvassed recently have wholeheartedly supported the selection of the Möbius loop as Canada's recycling symbol." The nation's manufacturers are expected to stamp the symbol on products containing reused materials. The logo breaks the strip into three fat, twisted arrows as shown below:

The number I gave for Lee De Forest's 1923 patent (it has been removed from the text) is incorrect, as several readers informed me, but I was unable to locate my source for this number, or to learn the correct one. I'll be pleased to hear from any reader who can run it down. In 1986 I was told that IBM was selling a printer that used a Möbius-strip ribbon to double the ribbon's life.

11. POLYHEXES AND POLYABOLOES

AT THE REQUEST of many readers, Figure 136 shows the 22 pentahexes. I know of no confirmed counts of distinct polyhexes (excluding reflections but including pieces with holes) beyond order 12, as mentioned in the chapter's addendum, or of any progress in finding an ennumeration formula.

Andrew Clarke investigated the tiling of polyaboloes. All shapes of orders 1 through 4 tile the plane. Clarke found that all but four of the pentaboloes tile, and all but nineteen of the hexaboloes. He also investigated 3-D analogs consisting of solids formed by joining half-cubes. The half-cubes, obtained by slicing unit cubes diagonally, are put together so at least one edge of one coincides with an edge of another, and so there is some area of surface contact. Three half-cubes joined in this manner, in all possible ways, produce twelve solids which can be used to produce solid figures in the manner of polycube pieces.

12. PERFECT, AMICABLE, SOCIABLE

THE CONJECTURE THAT there are no odd perfect numbers continues to be one of the most notorious unsolved problems in number theory. It could turn out to be undecidable, in which case it would have to be true. Why? Because if it were false there would be a counterexample (an odd perfect) and that would make the conjecture decidable.

There is a voluminous literature on the properties odd perfects would have if they existed. The lower bound was raised in 1976 to 10^{200}, and by now has perhaps gone higher. Euclid showed that an odd perfect must have the form $k(p^{4m+1})$, where p is an odd prime and k is a perfect square. However, not all num-

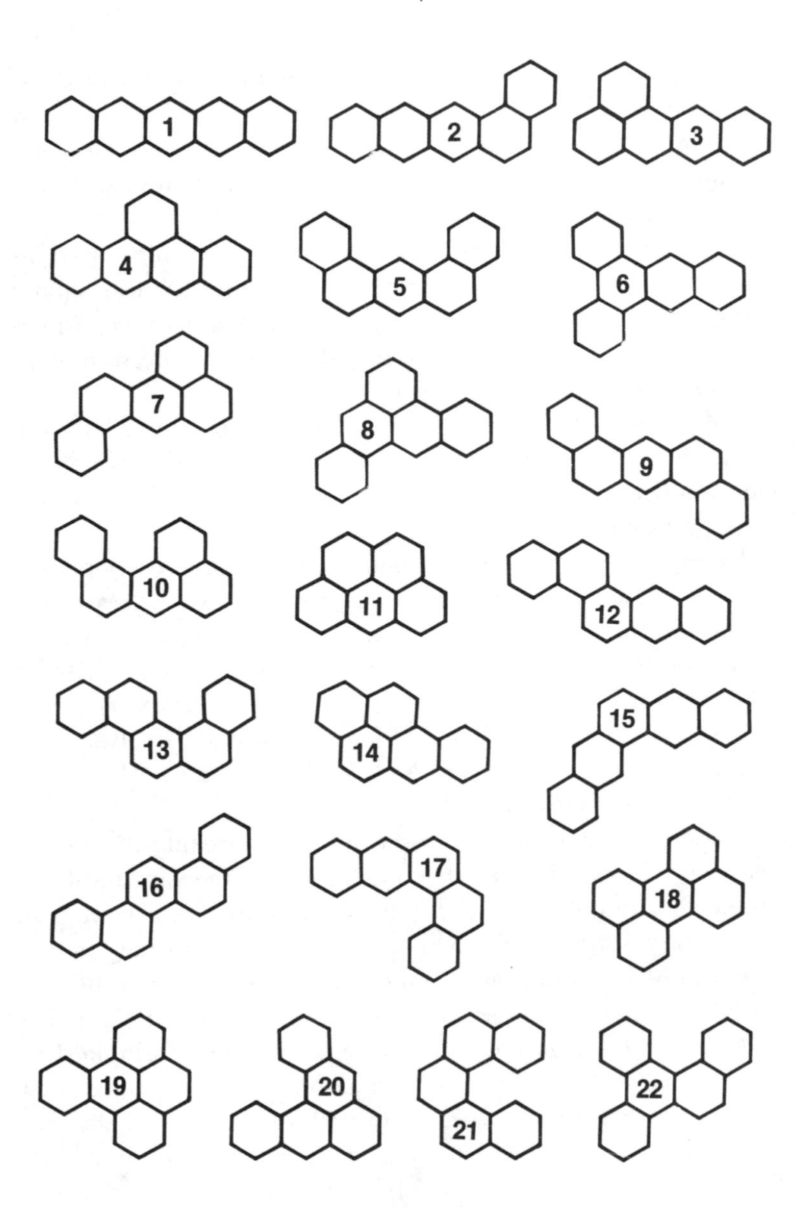

FIGURE 136
The 22 Pentahexes

bers of this form, 243 for example, are perfect. In 1980 it was shown that an odd perfect must have at least eight distinct factors.

Can an odd perfect be a square? A "no" answer is easily proved on the basis of the fact cited earlier that the sum of a perfect number's divisors is $2n$, an even number. If a number is odd, all of its divisors will be odd. If an integer is a square it has an odd number of divisors: its square root plus the evenly paired divisors less than its square root and those larger than its square root. Hence, an odd perfect number that was a square would have to have an odd number of odd divisors. But the sum of an odd number of odd numbers cannot be even.

Since this book's first edition in 1977, seven more Mersenne primes have been found by computer searches. This extends the table shown in Figure 71 to 31 perfect numbers. The new perfects are listed in Figure 137.

A curious, little known fact about perfects is that every perfect number (assuming there are no odd perfects) can be expressed by the form $6x + 28y$, where x and y are nonnegative integers. Note that the coefficients are the first two perfects. For example, 496 (the third perfect) is equal to $(6 \times 50) + (28 \times 7)$. For a simple proof see "Sum of Perfect Numbers," by Reinaldo Giudice, solution to problem 954, in *Mathematics Magazine*, 49, November 1976, page 257.

The past decade has witnessed a veritable explosion in the discovery of new amicable pairs as number theorists kept finding new formulas for them. Elvin Lee, of Fargo, North Dakota, one of the most active researchers in this area, tells me that the single most important achievement is a theorem presented by the German mathematician Walter Borho in his 1972 paper (cited in the bibliography) on Thabit ibn Kurrah's formula. Lee was the first to show how to obtain an unlimited number of new formulas from Borho's theorem. By 1989 more than 55,000 amicable pairs were known. The largest, found by Herman J. J. te Riele, has 282 digits. Lee informs me that a still larger pair, each exceeding 600 digits, has been found in Germany but I do not have the details. In an exhaustive search for amicables less than 10,000,000,000, te Riele found 1,427 pairs (see the bibliography for his report on this).

Many conjectures have been proved or disproved. One of the most interesting proofs, by Carl Pomerance, is that the sum of the reciprocals of all amicable numbers converges. The long-standing guess that every odd amicable is a multiple of 3 was shot down, as well as my second conjecture that the sum of every even amicable pair is equal to 0 or 7 (modulo 9). In 1984 te Riele found two counterexamples, the smallest of which is the pair 967947856 and 1031796176. Its sum equals 3 (modulo 9).

In 1988 two mathematicians reported the falsity of the conjecture that every odd amicable number is a multiple of 3. (See the last entry in the bibliography for amicable numbers.) The authors give fifteen counterexamples, of which the smallest is $a(140453)(85857199)$ and $a(56099)(214955207)$, where $a = 5^4 \times 7^3 \times 11^3 \times 13^2 \times 17^2 \times 19 \times 61^2 \times 97 \times 307$. Smaller counterexamples may exist. Whether there is an odd amicable pair with just one number divisible by 3 remains open.

The two outstanding questions about amicables remain unanswered. Is there a crowd? Is the set of amicable pairs finite or infinite?

Mersenne primes should not be confused with Fermat primes which have the form $2^n + 1$. Only five such primes are known ($n = 2^0, 2^1, 2^2, 2^3, 2^4$), and it is not known if their number is finite or infinite. Fermat proved that if such a number is prime then n is a power of two. He conjectured that numbers of this form are always prime when n is a power of 2, but his conjecture fails for $n = 32$. In 1988 the lowest then untested number of this form was $2^{2^{20}} - 1$, and it was found to be composite.

25	$2^{21700}(2^{21701} - 1)$	29	$2^{110502}(2^{110503} - 1)$
26	$2^{23208}(2^{23209} - 1)$	30	$2^{132048}(2^{132049} - 1)$
27	$2^{44496}(2^{44497} - 1)$	31	$2^{216090}(2^{216091} - 1)$
28	$2^{86242}(2^{86243} - 1)$		

FIGURE 137

Perfect numbers discovered since 1977. The numbers inside parentheses are Mersenne primes. The 29th Mersenne, discovered in 1988, turned out to be smaller than the next two which had been found earlier. Other Mersenne primes, and their corresponding perfects, may still be lurking in the gaps.

13. POLYOMINOES AND RECTIFICATION

THE UNSOLVED RECTIFICATION problems mentioned in the chapter, involving the hexomino and the heptomino shown at the right of Figure 81, were both solved in 1987 by Karl Dahlke, a software engineer at AT&T Bell Laboratories, in Naperville, Illinois. Dahlke is blind, but he has a personal computer with a speech synthesizer that translates the machine's output into sound.

At first Dahlke tried to prove that both rectifications were impossible. Failing at this, he began a systematic computer search for the smallest rectangle that perhaps could be formed by replicas of each piece. His success was reported by Ivars Peterson in *Science News*, Vol. 132, November 14, 1987, page 132.

Solomon Golomb worked on both problems. He found that each piece would tile an infinite half-strip (one that goes to infinity in only one direction). He also found that each piece will tile a rectangle with a unit hole. "I was amazed," he told *Science News* when informed of Dahlke's solutions. "A lot of very bright people have worked on [the problems]. It is a noteworthy accomplishment."

Actually, credit for the first rectification of the heptomino should go to Karl Scherer, of Auckland, New Zealand. He proposed it as a problem in the *Journal of Recreational Mathematics*, Vol. 14, No. 1, 1981/82, p. 64. Scherer, who had found a rectification, gave the rectangle's size as 16 × 42, larger than Dahlke's 21 × 26. Because no reader solved the problem, publication of Scherer's solution was delayed. See the journal's Vol. 21, No. 3, 1989.

Dahlke contributed two short papers to the *Journal of Combinatorial Theory*, Series A, May 1989. The paper titled "The *Y*-Hexomino Has Order 92" (pp. 125–126), asserts that the smallest rectangle that can be tiled with the hexomino is 23 × 24. It contains 92 hexominoes, the largest number for a minimal tiling of any hexomino. The other paper, titled "A Hexomino of Order 76" (pp. 127–128), reported that the 19 × 28 rectangle is the smallest solution for the heptomino. It contains 76 replicas, or two less than the number in the solution Dahlke found earlier and which was published in *Science News*. The two minimal solutions are shown in Figure 138.

FIGURE 138

Karl Dahlke's solutions to two difficult rectification tasks. Above, the 92 copies of the hexomino tile, a 23 × 24 rectangle obtained in 1987. On the bottom, 76 copies of a heptomino tile, a 19 × 28 rectangle worked out in 1988.

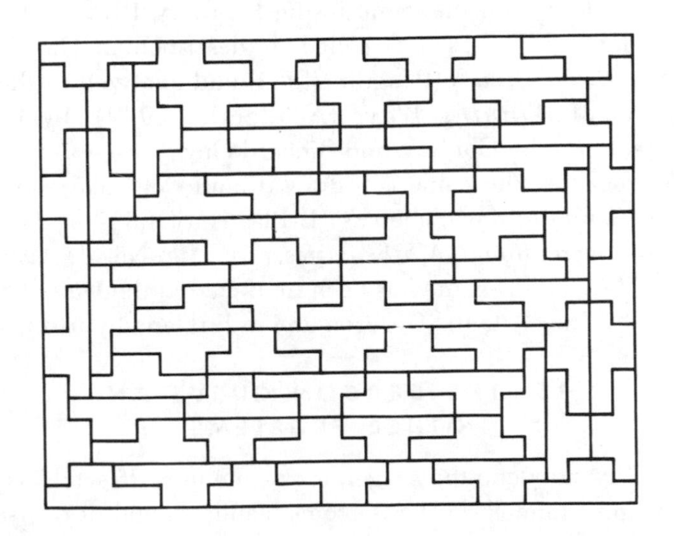

In a paper on "Polyominoes Which Tile Rectangles," in the *Journal of Combinatorial Theory*, Series A (May 1989), pages 117–124, Solomon Golomb reports on some results involving what I shall call the replication order (RO) of a polyomino. This is the smallest number of replications that will form a rectangle. A polyomino has RO 1 if and only if it is itself a rectangle. (RO is undefined for a polyomino that cannot tile a rectangle.) Golomb describes the conditions for a polyomino to have RO 2 and 4, and shows that there are infinitely many polyominoes of RO 2, and of all ROs that are multiples of 4.

Golomb lists many unsolved problems. For example, single instances are known for ROs 10 and 18. Is there a polyomino for every even RO? The chief unsolved question is whether there is a polyomino, aside from the trivial RO 1, with an odd RO. Small odd ROs such as 3 and 5 "seem particularly unlikely," Golomb writes, but he sees no reason why larger odd replication orders are not possible.

Edward de Bono invented and patented a simple but elegant little two-person game played on a 4 × 4 field with two L-tetrominoes and two monominoes (unit squares). You'll find the rules explained in his Pelican paperback, *The Five-Day Course in Thinking*. He also wrote about the game in the British monthly, *Games and Puzzles* (November 1974), pages 4–6 (see also letters on the game in the February 1975 issue, page 36). The L-game, as it is called, is described in David Pritchard's *Brain Games* (Penguin, 1982) and analyzed in the first volume of *Winning Ways* (Academic, 1982), by Elwyn Berlekamp, John Conway, and Richard Guy.

A proof that the game is a draw if both sides play rationally will be found in Karl Scherer's "L-Play is a Draw," in the *Journal of Recreational Mathematics*, Vol. 12, No. 1, 1979–80, pages 2–8. The L-game has been marketed in the United States by JABO, Inc., Atlanta, Georgia, and in England by Just Games, London.

15. THE DRAGON CURVE AND OTHER PROBLEMS

SINCE I INTRODUCED the dragon curve, the term "fractal," coined by Benoit Mandelbrot, has become standard, and of course the dragon curve is a fractal. Books on fractals, including Mandelbrot's classic *Fractal Geometry of Nature* (Freeman, 1982), are appearing so rapidly that I will make no effort to list any here. A selective bibliography can be found at the end of Chapter 3, "Mandelbrot's Fractals," in my *Penrose Tiles to Trapdoor Ciphers* (Freeman, 1988). For fascinating generalizations of the dragon curve to three dimensions, see "Wire Bending," by Michel Mendes France and J. O. Shallit, in the *Journal of Combinatorial Theory*, Series A, Vol. 50, January 1989, pages 1–23.

The *Scientific American* column on gray codes, cited in the answer to Problem 5, is reprinted in my *Knotted Doughnuts and Other Mathematical Entertainments* (Freeman, 1986).

16. COLORED TRIANGLES AND CUBES

A Tokyo correspondent informed me in 1974 that the Diophantine equation at the top of page 234 had been studied by Professor Uchiyama (I do not know his first name). Uchiyama reportedly showed that beyond $m = 24$ there are at most two solutions. He conjectures that actually there are none.

I mentioned that I had earlier written about MacMahon's set of 30 color cubes in my *New Mathematical Diversions from Scientific American*, now available in paper covers from the University of Chicago Press. I returned to the 30 cubes in my September 1978 column, not yet reprinted in a book, where I give some elegant new results discovered by John H. Conway.

17. TREES

Isaac Asimov, who obtained his doctorate in biochemistry, wrote to tell me how neatly the trees depicted in Figure 119 count hydrocarbon isomers. (Isomers are compounds with the same number of atoms of each element, but with the atoms differently linked.) Open carbon chains are trees in which no point can be connected to more than four others. The unique two-point tree corresponds to ethane; the unique three-point tree to propane. The two four-point trees give butane and its isomer isobutane. The three five-point trees provide three isomers: pentane, isopentane, and neopentane. The first five of the six six-point trees correspond to normal hexane, 2-methyl pentane, 3-methyl pentane, 2,3-dimethyl butane, and 2,2-dimethyl butane. Because the sixth tree has a point joined to five others, it corresponds to no hydrocarbon.

Hydrocarbon molecules can link to form rings "to further complicate matters," as Asimov put it, "and amuse the recreational mathematician." Asimov wonders if graph theory made it possible for chemists to decide that a molecule with forty carbon atoms and eighty-two hydrogen atoms has precisely 62,491,178,805,831 isomers.

18. DICE

FOR SOME MORE startling curiosities about dice, see the discussion of nontransitive dice in Chapter 5 of my *Wheels, Life, and Other Mathematical Diversions* (Freeman, 1983), and Sicherman's dice in Chapter 19 of my *Penrose Tiles to Trapdoor Ciphers* (Freeman, 1989).

19. EVERYTHING

IT IS SURELY APPARENT from my chapter on everything that I am an unabashed Platonist in the following sense. I believe that both the physical world and the abstract world of pure mathematics have an existence that is not dependent on the existence of human beings. If any readers are interested in my arguments for this utterly commonplace view, held by everybody except a small number of thinkers smitten by the notion that humanity is the measure of all things, they can consult the first chapter of my *Whys of a Philosophical Scrivener*; Part 1, Chapter 5; Part 2, Chapter 34, of my *Order and Surprise*; and my Guest Comment in the *American Journal of Physics*, April 1989, page 203.

Bibliography

1. NOTHING

"Oom." Martin Gardner. *The Journal of Science Fiction*, Fall 1951. Reprinted in Martin Gardner, *The No-Sided Professor and Other Tales*, Prometheus, 1987.

The Annotated Snark. Martin Gardner. Simon and Schuster, 1962.

Nil: Episodes in the Literary Conquest of Void During the Nineteenth Century. Robert Martin Adams. Oxford University Press, 1966.

"Nothing." P. L. Heath in *The Encyclopedia of Philosophy*. Macmillan-Free Press, 1967.

"Why." Paul Edwards in *The Encyclopedia of Philosophy*. Macmillan-Free Press, 1967.

All Numbers Great and Small. J. H. Conway. University of Calgary Mathematical Research Paper No. 149, February 1972.

Surreal Numbers. Donald E. Knuth. Addison-Wesley, 1974.

On Numbers and Games. J. H. Conway. Academic Press, 1976.

"Conway's Surreal Numbers." Martin Gardner in *Penrose Tiles to Trapdoor Ciphers*. W. H. Freeman, 1989, Chapter 4.

3. GAME THEORY, GUESS IT, FOXHOLES

Strategy in Poker, Business and War. John D. McDonald. W. W. Norton & Company, 1950.

Introduction to the Theory of Games. J. C. C. McKinsey. McGraw-Hill, 1952.

The Compleat Strategyst: Being a Primer on the Theory of Games of Strategy. J. D. Williams. McGraw-Hill, 1954.

Theory of Games as a Tool for the Moral Philosopher. R. B. Braithwaite. Cambridge University Press, 1955.

The Theory of Games and Linear Programming. S. Vadja. John Wiley & Sons, 1957.

Games and Decisions. R. Duncan Luce and Howard Raiffa (eds.). John Wiley, 1957.

Introduction to the Theory of Games. E. Burger. Prentice-Hall, 1963.

Differential Games. Rufus Isaacs. John Wiley, 1965.

Two-Person Game Theory: The Essential Ideas. Anatol Rapoport. University of Michigan Press, 1970.
Differential Games. Avney Friedman. Wiley-Interscience, 1971.
Differential Games and Related Topics. H. W. Kuhn and G. P. Szegö (eds.). North-Holland, 1971.

4. FACTORIAL ODDITIES

"Leonhard Euler's Integral: A Historical Profile of the Gamma Function." P. J. Davis. *American Mathematical Monthly,* Vol. 66, December 1959, pages 849–69. A history of the extension of factorials to nonintegers.
Recreations in the Theory of Numbers. Albert H. Beiler. Dover, 1964, Chapter 7.
Mathematics of Choice. Ivan Niven, New Mathematical Library, Random House, 1965.
"Exclamation Point!" Isaac Asimov in *From Earth to Heaven.* Avon, 1966, Chapter 7.
"The Digits of a Factorial." Charles W. Trigg. *Mathematics Magazine,* Vol. 40, May 1967, page 165.
"A Factorial Conjecture." Gustavus J. Simmons. *Journal of Recreational Mathematics,* Vol. 1, January 1968, page 38; Vol. 3, October 1970, page 232.
Fundamental Algorithms. Donald E. Knuth. Addison-Wesley, 1968, pages 44–51.
"A Note on N!" J. M. Maxfield. *Mathematics Magazine,* Vol. 43, March 1970, pages 64–67.
"Rate Your Wits! (Factorials)." David L. Silverman. *Journal of Recreational Mathematics,* Vol. 3, July 1970, pages 174–175.
"Integers and the Sum of the Factorials of Their Digits." George D. Poole. *Mathematics Magazine,* Vol. 44, November 1971, pages 278–279.
"Sums of Factorials." Victor G. Feser. *Journal of Recreational Mathematics,* Vol. 5, July 1972, pages 174–176.
"On the Distribution of Decimal Digits in *n*!" S. P. Castell. *Eureka,* No. 36, October 1973, pages 44–47.
"Investigation of Maxfield's Theorem." Laura Southard. *Pi Mu Epsilon Journal,* Vol. 7, Spring 1983, pages 493–495.
"Factorial!" Robert Messer. *Mathematics Teacher,* Vol. 77, January 1984, pages 50–51.

6. DOUBLE ACROSTICS

Excursions into Puzzledom. "Tom Hood and His Sister." London: Strahan, 1879.
Acrostic Dictionary. Phillippa Pearson. London, 1884.
The Illustrated Book of Puzzles. Don Lemon (editor). London: Saxon, 1892.
Everybody's Illustrated Book of Puzzles. "The Sphinx." London: Brindley & Howe, no date.
A Book of Acrostics. Ronald Arbuthnott Knox. London: Methuen, 1924.
The Strand Problems Book. W. T. Williams and G. H. Savage. London: George Newnes, no date, pages 79–82.

The World's Best Word Puzzles. Henry Ernest Dudeney. London: Daily News, 1925. Reprinted as *300 Best Word Puzzles*, Charles Scribner's Sons, 1968.

Acrostic Dictionary and Crossword Companion. W. M. Baker. London, 1934.

Hubert Phillips's Heptameron. Hubert Phillips. London: Eyre & Spottiswoode, 1945.

"15 Letters: Most Popular Game." John M. Willig. *The New York Times Magazine*, December 15, 1963. History of the crossword.

"Crossword Puzzle." Margaret Farrar in *Encyclopedia Americana*. Americana Corp., current edition.

The Strange World of the Crossword. Roger Millington. London: M. & J. Hobbs in association with Michael Joseph, Ltd., 1974.

"The First Double and Triple Acrostics." W. Ross Eckler. *Word Ways*, Vol. 19, November 1986, pages 228–229.

7. PLAYING CARDS

Mathematics, Magic, and Mystery. Martin Gardner. Dover, 1956.

Mathematical Magic. William Simon. Scribner, 1964.

Scarne on Card Tricks. John Scarne. Crown, 1974.

Self-Working Card Tricks. Karl Fulves. Dover, 1976.

"Mathematical Tricks with Cards." Martin Gardner in *Wheels, Life, and Other Mathematical Amusements*. Freeman, 1983, Chapter 19.

8. FINGER ARITHMETIC

"Digital Reckoning Among the Ancients." Leon J. Richardson. *American Mathematical Monthly*, Vol. 23, January 1916, pages 7–13.

"Mathematics and the Folkways." Martin Gardner. *Journal of Philosophy*, Vol. 47, March 30, 1950, pages 177–86.

History of Mathematics, Vol. 2. David Eugene Smith. Dover reprint, 1958, pages 119–20, 196–202.

"Digital Computer—Nonelectronic." Ferd W. McElwain. *Mathematics Teacher*, Vol. 54, April 1961, pages 224–28.

"Finger Reckoning." Robert W. Prielipp. *Mathematics Teacher*, Vol. 61, January 1968, pages 42–43.

"Finger Reckoning in an Arabian Poem." A. S. Saidan. *Mathematics Teacher*, Vol. 61, November 1968, pages 707–08.

Number Words and Number Symbols. Karl Menninger. M.I.T. Press, 1969, pages 201–20.

Seminumerical Algorithms. Donald E. Knuth. Addison-Wesley, 1969, Chapter 4.

The Complete Book of Fingermath. Edward M. Lieberthal. McGraw-Hill, 1979.

9. MÖBIUS BANDS

"Topological Tomfoolery." *Mathematics, Magic, and Mystery*. Martin Gardner. Dover 1956, Chapter 5.

"Curious Topological Models." *The Scientific American Book of Mathematical Puzzles and Diversions.* Martin Gardner. Simon and Schuster, 1959, Chapter 7; University of Chicago Press, revised edition, 1988, with new title of *Hexaflexagons and Other Mathematical Diversions.*

Intuitive Concepts in Elementary Topology. B. H. Arnold. Prentice-Hall, 1962.

"Chemical Topology." Edel Wasserman. *Scientific American,* November 1962, pages 94–102.

Experiments in Topology. Stephen Barr. Thomas Y. Crowell, 1964.

Visual Topology. W. Lietzmann. American Elsevier Publishing Company, 1965.

"Klein Bottles and Other Surfaces." *The Sixth Book of Mathematical Games from Scientific American.* Martin Gardner. Freeman, 1971, Chapter 2; revised edition, University of Chicago Press, 1971.

"The Möbius Band in Roman Mosaics." Lorraine L. Larison. *American Scientist,* Vol. 61, September-October, 1973, pages 544–547.

"No-Sided Professor." Martin Gardner in *The No-Sided Professor and Other Tales.* Prometheus, 1987.

11. POLYHEXES AND POLYABOLOES

"Some Tetrabolical Difficulties." Thomas H. O'Beirne. *New Scientist,* Vol. 13, January 18, 1962, pages 158–59.

Polyominoes. Solomon W. Golomb. Charles Scribner's Sons, 1965.

"Counting Hexagonal and Triangular Polyominoes." W. F. Lunnon in *Graph Theory and Computing.* Academic Press, 1972.

12. PERFECT, AMICABLE, SOCIABLE

Perfect Numbers

"Perfect Numbers in the Binary System." Evelyn Rosenthal. *Mathematics Teacher,* April 1962, pages 249–260.

"Comments on the Properties of Odd Perfect Numbers." Lee Ratzan. *Pi Mu Epsilon Journal,* Spring, 1972, pages 265–271.

Perfect Numbers. Richard W. Shoemaker. National Council of Teachers of Mathematics, 1973.

"Almost Perfect Numbers." R. P. Jerrard and Nicholas Temperley. *Mathematics Magazine,* Vol. 46, March 1973, pages 84–87.

"A Search Procedure and Lower Bound for Odd Perfect Numbers." Bryant Tuckerman. *Mathematics of Computation,* Vol. 27, October 1973, pages 943–49.

"A Note on Almost Perfect Numbers." James T. Cross. *Mathematics Magazine,* Vol. 47, September 1974, pages 230–31.

Perfect Numbers. Stanley Bezuszka, Margaret Kenney, and Stephen Kokoska. Boston College Press, 1980.

"Perfect Numbers." Stan Wagon. *The Mathematical Intelligencer,* Vol. 7, No. 2, 1985, pages 66–68.

Amicable Numbers

"Amicable Numbers." Edward Brind Escott. *Scripta Mathematica,* Vol. 12, March 1946, pages 61–72.

Excursions into Mathematics. Anatole Beck, Michael N. Bleicher, and Donald W. Crowe. Worth, 1962, Chapter 2.

"Friendly Numbers." Howard L. Rolf. *Mathematics Teacher,* Vol. 60, February 1967, pages 157–60.

"Amicable Numbers with Opposite Parity." A. A. Gioia, and A. M. Vaidya. *The American Mathematical Monthly,* Vol. 74, October 1967, pages 969–973.

"Amicable Numbers and the Bilinear Diophantine Equation." Elvin J. Lee. *Mathematics of Computation,* Vol. 22, January 1968, pages 181–187.

"On Division by Nine of the Sums of Even Amicable Pairs." Elvin J. Lee. *Mathematics of Computation,* Vol. 23, July 1969, pages 545–548.

"Unitary Amicable Numbers." Peter Hagis, Jr. *Mathematics of Computation,* Vol. 25, October 1971, pages 915–918.

"Relatively Prime Amicable Numbers with Twenty-One Prime Divisors." Peter Hagis, Jr. *Mathematics Magazine,* Vol. 45, January 1972, pages 21–26.

"On Thabit ibn Kurrah's Formula for Amicable Numbers." Walter Borho. *Mathematics of Computation,* Vol. 26, April 1972, pages 303–304.

"The History and Discovery of Amicable Numbers." Elvin J. Lee and Joseph Madachy. *Journal of Recreation Mathematics,* Vol. 5, Nos. 2, 3, 4, 1972. The first paper lists 67 references.

"On Generating New Amicable Pairs from Given Amicable Pairs." Herman J. J. te Riele. *Mathematics of Computation,* Vol. 42, January 1984, pages 219–223.

"New Unitary Amicable Couples." Mariano Garcia. *Journal of Recreational Mathematics,* Vol. 17, No. 1, 1984–85, pages 32–35.

"Computation of All the Amicable Pairs Below Ten Billion." Herman J. J. te Riele. *Department of Numerical Mathematics Report NM-R8503,* Center for Mathematics and Computer Science, Amsterdam, March 1985.

"Table of Amicable Pairs Between 10^{10} and 10^{52}." Herman J. J. te Riele, Walter Borho, S. Battiato, H. Hoffmann, and Elvin J. Lee. *Department of Numerical Mathematics Note NM-N8603,* Center for Mathematics and Computer Science, Amsterdam, September 1986.

"Are There Odd Amicable Numbers Not Divisible by Three?" S. Battiato and W. Borho. *Mathematics of Computation,* Vol. 50, April 1988, pages 633–637.

Sociable Numbers

"On Amicable and Sociable Numbers." Henri Cohen. *Mathematics of Computation,* Vol. 24, April 1970, pages 423–429.

13. POLYOMINOES AND RECTIFICATION

Pentomino games

"Polyominoes." *The Scientific American Book of Mathematical Puzzles and Diversions.* Martin Gardner. Simon and Schuster, 1959, Chapter 13; University of Chicago Press, revised edition, 1988 with new title of *Hexaflexagons and Other Mathematical Diversions.*

Polyominoes. Solomon W. Golomb. Charles Scribner's Sons, 1965.

Brain Games. David Pritchard. Penguin, 1982.

Rectification

"Covering a Rectangle with *L*-tetrominoes." *American Mathematical Monthly,* Vol. 70, August 1963, pages 760–61. David A. Klarner's solution to S. W. Golomb's problem E1543.

"Covering a Rectangle with *T*-tetrominoes." D. W. Walkup. *American Mathematical Monthly,* Vol. 72, November 1965, pages 986–88. It is shown that a necessary and sufficient condition for an $a \times b$ rectangle to be covered by *T*-tetrominoes is that a and b be integral multiples of 4. This solves Robert Spira's problem E1786, which appeared in the same magazine, May 1965.

"Some Results Concerning Polyominoes." David Klarner. *Fibonacci Quarterly,* Vol. 3, February 1965, pages 9–20.

"Tiling with Polyominoes." S. W. Golomb. *Journal of Combinatorial Theory,* Vol. 1, September 1966, pages 280–96. Includes the rectification problem for polyominoes through order 6.

"Impossibility of Covering a Rectangle with *L*-hexominoes." *American Mathematical Monthly,* Vol. 75, August 1968, pages 785–86. Solution by Dennis Gannon to Robert Spira's problem E1983.

"Packing a Rectangle with Congruent *N*-ominoes." David A. Klarner. *Journal of Combinatorial Theory,* Vol. 7, September 1969, pages 107–15. The latest, most complete discussion of rectification, including many unsolved problems.

"Tiling $5n \times 12$ Rectangles with *Y*-Pentominoes." James Bitner. *Journal of Recreational Mathematics,* Vol. 7, Fall 1974, pages 276–278.

"Tiling Rectangles with *T* and *C* Pentominoes." Earl S. Kramer. *Journal of Recreational Mathematics,* Vol. 16, No. 2, 1983–84, pages 102–113.

14. KNIGHTS OF THE SQUARE TABLE

Mathematical Recreations and Essays. W. W. Rouse Ball. Macmillan, 1892; thirteenth revised edition, Dover, 1987, Chapter 6.

Games Ancient and Oriental. Edward Falkener. Longmans, Green, 1892, Dover, 1961, Chapters 38, 39, 40, 42.

Mathematische Unterhaltungen und Spiele. W. Ahrens. Leipzig, 1910, Vol. 1, Chapter 11; 1918, Vol. 2, pages 354–60.

Amusements in Mathematics. H. E. Dudeney. Nelson, 1917, Dover, 1958, pages 101–03, 127.

Le Problème du Cavalier. Maurice Kraitchik. Brussels, 1927.

Le Problème du Cavalier Généralisé. E. Huber-Stockar. Brussels, 1935.

Mathematical Recreations. Maurice Kraitchik. W. W. Norton & Company, 1942, Dover, 1953, Chapter 11.

Les Secrets du Cavalier. G. D'Hooghe. Brussels, 1962.

"Magic Knight Tours on Square Boards." T. H. Willcocks. *Recreational Mathematics Magazine,* No. 12, December 1962, pages 9–13.

"The Construction of Magic Knight Tours." T. H. Willcocks. *Journal of Recreational Mathematics,* Vol. 1, October 1968, pages 225–33.

"Of Knights and Cooks, and the Game of Cheskers." Solomon W. Golomb. *Journal of Recreational Mathematics,* Vol. 1, July 1968, pages 130–38.

"Knight's Tour Revisited." Paul Cull and Jeffery De Curtins. *The Fibonacci Quarterly,* Vol. 16, June 1978, pages 276–286.

16. COLORED TRIANGLES AND CUBES

New Mathematical Pastimes. Major P. A. MacMahon. Cambridge University Press, 1921.
"Domino and Superdomino Recreations, Part 6." Wade E. Philpott. *Journal of Recreational Mathematics,* Vol. 6, Winter 1973, pages 10–34.
"Patterns in Space." Marc Odier. *Games & Puzzles,* No. 37, June 1975, pages 11–17.
Surprenants Triangles. Marc Odier and Y. Roussel. CEDIC, 1976.

17. TREES

The Theory of Graphs and Applications. Claude Berge. John Wiley & Sons, 1962.
Graphs and Their Uses. Øystein Ore. Random House, 1963.
Finite Graphs and Networks: An Introduction with Applications. Robert G. Busacker and Thomas L. Saaty. McGraw-Hill, 1965.
"Trees and Unicyclic Graphs." Sabra Anderson and Frank Harary. *Mathematics Teacher,* April 1967, pages 345–348.
Graph Theory. Frank Harary. Addison-Wesley, 1969.
Introduction to Graph Theory. Robin Wilson. Academic, 1972.
"Trees." Chapter 3, *Graph Theory 1736–1936.* Norman Biggs, Keith Lloyd, and Robin Wilson (eds.). Oxford University Press, 1976.
Trees. Jean-Pierre Serre. Springer-Verlag, 1980.
Trees and Applications. Frank Harary and Michael Plantholt. In preparation.

18. DICE

Scarne on Dice. John Scarne and Clayton Rawson. Harrisburg, Pa.: Military Service Publishing Co., 1945; ninth revised edition, 1968.
How to Figure the Odds. Oswald Jacoby. Doubleday, 1947.
Cardano, the Gambling Scholar. Øystein Ore. Princeton University Press, 1953, Dover, 1965.
Scarne's Complete Guide to Gambling. John Scarne. Simon and Schuster, 1961.
Marked Cards and Loaded Dice. Frank Garcia. Prentice-Hall, 1962.
"Repeated Independent Trials and a Class of Dice Problems." Edward O. Thorp. *American Mathematical Monthly,* Vol. 71, August 1964, pages 778–81.
The Casino Gambler's Guide. Allan N. Wilson. Harper & Row, 1965.
The Theory of Games and Statistical Logic. Richard A. Epstein. Academic Press, 1967, Chapter 5.
Casino Holiday. Jacques Noir. Oxford Street Press, 1968.
Dice: Squares, Tops, and Shapes. Burton Williams. GBC Press, Las Vegas, 1982.
Gambling Scams. Darwin Ortiz. Dodd, Mead, 1984.

19. EVERYTHING

"On What There Is." Willard Van Orman Quine in *From a Logical Point of View.* Harvard University Press, 1953, revised 1961.

"Existence." A. N. Prior in *The Encyclopedia of Philosophy.* Macmillan-Free Press, 1967.

"Superspace." John Archibald Wheeler in *Analytic Methods in Mathematical Physics,* edited by Robert P. Gilbert and Roger G. Newton. Gordon and Breach, 1970.

"Beyond the Black Hole." John Archibald Wheeler in *Science Year: 1973.* Field Enterprises, 1972.

Gravitation. Charles W. Misner, Kip S. Thorne, and John Archibald Wheeler. W. H. Freeman, 1973.

"Is Physics Legislated by Cosmogony?" Charles M. Patton and John Archibald Wheeler in *Quantum Gravity: An Oxford Symposium,* edited by C. J. Isham, Roger Penrose, and D. W. Sciama. Clarendon Press, 1975.

"Genesis and Observership." John Archibald Wheeler. Joseph Henry Laboratories, Princeton University, March 26, 1976, preprint of unpublished paper.